A Primer on Machine Learning Applications in Civil Engineering

A Primer on Machine Learning Applications in Civil Engineering

Paresh Chandra Deka

CRC Press
Taylor & Francis Group
Boca Raton London New York

CRC Press is an imprint of the
Taylor & Francis Group, an **informa** business

CRC Press
Taylor & Francis Group
6000 Broken Sound Parkway NW, Suite 300
Boca Raton, FL 33487-2742

© 2020 by Taylor & Francis Group, LLC
CRC Press is an imprint of Taylor & Francis Group, an Informa business

No claim to original U.S. Government works

Printed on acid-free paper

International Standard Book Number-13: 978-1-138-32339-1 (Hardback)

To my wife, Rumi, and daughter, Fairyqueen

Contents

Preface

A Primer on Machine Learning Applications in Civil Engineering delivers the main contemporary themes and tools in machine learning, including artificial neural networks, fuzzy logic, support vector machines, genetic algorithm, and hybrid systems. These themes are discussed or reformulated from either a local view or a global view. Different from previous books that only explain the mathematics of machine learning algorithms, this book presents a unified and new picture by discussing the applications of machine learning in the civil engineering domain. Within the new picture, different machine learning models and theories are bridged in an elegant and systematic manner. For precise and thorough understanding, this book also presents applications of the new hybrid algorithms along with data statistics.

This book not only provides researchers with the latest research results lively and timely, but also presents an excellent overview on machine learning. Importantly, the applications of a new line of learning through various models is made understandable to a large proportion of audience including researchers in machine learning and graduate students. To provide a comprehensive representation of research, this title has focused on applications in the field of structural engineering, transportation engineering, geotechnical engineering, construction engineering and management, coastal or marine engineering, environmental engineering, and water resources engineering via hydrological modeling.

MATLAB® is a registered trademark of The MathWorks, Inc. For product information, please contact:

The MathWorks, Inc.
3 Apple Hill Drive
Natick, MA, 01760-2098 USA
Tel: 508-647-7000
Fax: 508-647-7001
E-mail: info@mathworks.com
Web: www.mathworks.com

Acknowledgments

It is my pleasure to express profound gratitude and indebtedness towards my research supervisor Prof V Chandramouli as well as my students throughout the world for their continued inspiration, motivation, support, discussions, and great patience towards research, which made this book possible.

I am also grateful to NITK and all other faculty members and staff of the Department of Applied Mechanics and Hydraulics, NITK, Surathkal, for helping me directly or indirectly during my work. The blessings of my father and mother are highly acknowledged towards the commitment of work

Without the support, patience, and encouragement from my wife, Rumi, and my daughter, Fairyqueen, I would never have been able to submit this work.

Paresh Chandra Deka
NITK, Surathkal

A Primer on Machine Learning Applications in Civil Engineering

The term 'machine learning,' sometimes referred to as computational intelligence, is a multidisciplinary field which includes a set of soft computing techniques and algorithms that deal with complex natural systems, exploiting the given tolerance for imprecision and uncertainty for a particular problem with backup from mathematical and computational approaches. Beginning with an overview of machine learning, the book offers exhaustive coverage of artificial neural networks (ANNs), discussing in detail their fundamental concepts, evolution, and recent developments. Various learning algorithms under the supervised and unsupervised categories, along with special networks, are included. This is followed by detailed chapters discussing fuzzy logic features and applications. A detailed explanation is given for fuzzy sets, membership functions, rule-based systems, and fuzzy logic control systems. In addition, the chapter on support vector machine (SVM) and its variants deals with various aspects of statistical learning theory and its advantages are incorporated. Various SVM classifications, kernel-based methods, and support vector regression are elaborately discussed. The chapter on genetic algorithms (GA) explains various GA operators, such as crossover and mutation, with suitable examples and illustrations. Different classes of GA are explained in detail. Genetic programming is discussed in brief. Hybrid intelligent systems are illustrated in detail; the engineering applications of swarm intelligent systems are discussed; a brief explanation about data analysis and data presentation, which is an integral part of machine learning, is included. Finally, the applications in various civil engineering domains have been explained with case studies.

This unique book delivers the main contemporary themes and tools in machine learning, including probabilistic generative models and support vector machines. These themes are discussed or reformulated from either a local view or a global view. Different from previous books that only explain the mathematics of machine learning algorithms, this book presents a unified and new picture by discussing the applications of machine learning in the civil engineering domain. State-of-the-art algorithms and methods for neural and fuzzy computing, support vector networks, and evolutionary computation are presented in a lucid manner with connection to specific civil engineering applications. The problems encountered in the civil engineering domain are generally unstructured and imprecisely influenced by the intuitions and past experiences of a designer. The conventional methods of computing that rely on analytical or empirical relations are time-consuming and labor-intensive when posed with real-time problems. To study, model, and analyze such problems, appropriate computer-based machine learning techniques inspired by human reasoning, intuition, consciousness, and wisdom are much more reliable and relevant. Within this new picture of different machine learning models and their theories, graduate students, researchers, and practitioners are highly benefitted. The main research trends are also pointed out at the end of each case study. For precise and thorough understanding, this book also presents applications of the new hybrid algorithms.

This book not only provides researchers with the latest research results, but also presents an excellent overview of machine learning approaches. Importantly, the applications of a new line of learning through various models are made understandable to a large proportion of the audience, including researchers in machine learning. This book can help inexperienced users solve engineering problems and experienced users improve work efficiency. To provide a comprehensive representation of research, this book has focused on applications in the field of geotechnical engineering, construction engineering, and management, coastal or marine engineering, water resources engineering via hydrological modeling, environmental engineering, structural engineering and transportation engineering. MATLAB codes have been used and incorporated to illustrate the applications of the concepts discussed. It also includes tutorials and new algorithms that describe and demonstrate all the previously mentioned techniques.

At the end, the future direction of research and developments, along with an overall conclusion about machine learning, is included.

Author

Paresh Chandra Deka earned a bachelor's in civil engineering at the National Institute of Technology, Silchar, Assam, India, and a PhD at the Indian Institute of Technology, Guwahati, specializing in hydrological modeling. Dr. Deka served on the faculty at the School of Postgraduate Studies at Arbaminch University, Ethiopia from 2005 to 2008 and as visiting faculty in 2012 at the Asian Institute of Technology, Bangkok, Thailand. He has supervised 10 PhD scholars as well as 5 current PhD scholars. He has supervised 40 master's theses as well as 4 current master's students. His research area is soft computing applications in water resources engineering and management.

Dr. Deka has published 4 books, 5 book chapters, and more than 40 international journal papers. He is a visiting faculty member doing short-term research interaction at Purdue University, Indiana. With more than 28 years of teaching experience, he is currently a professor in the Department of Applied Mechanics and Hydraulics at the National Institute of Technology, Surathkal, Karnataka, India.

1

Introduction

1.1 Machine Learning

Machine learning is related to artificial learning and has been getting tremendous attention from researchers across the globe since its inception. It is indeed a very powerful approach to data analysis and modeling for various applications. The fundamental feature of machine learning algorithms is that they can learn from empirical data and can be used where modeled phenomena are hidden, non-evident, or not very well explained. There are many different algorithms available in machine learning using many methods of nonparametric statistics, artificial intelligence research, and computer science. Here, the most commonly used MLAs for Civil Engineering related problems are Artificial Neural Network (ANN), Fuzzy Logic (FL), Genetic Algorithm (GA), and Support Vector Machine (SVM).

In the context of machine learning, ANN, FL, GA, and SVM are important models. These are not only approaches to develop artificial intelligence but are also used as universal nonlinear and adaptive tools for solving data-driven classification and regression problems. In this regard, machine learning is seen as an applied scientific discipline, while the general properties of statistical learning from data and the mathematical theory of generalization from experience are more relevant.

The complexity necessary to mathematically model real-world problems has compelled human civilization to search for nature-inspired computing tools. The evolution of such computing tools revolves around the information processing characteristics of biological systems. In contrast to conventional computing, these tools are rather 'soft' as they lack exactness and are, therefore, placed under the umbrella of a multidisciplinary field called soft computing.

Soft computing is an emerging collection of methodologies which aims to exploit tolerance for imprecision, uncertainty, and partial truth to achieve robustness, tractability, and cost savings. It is a branch of computational intelligence research that employs a variety of statistical, probabilistic, and optimization tools to learn from past examples and to then use that prior training to classify new data, identify new patterns, or predict novel trends.

Soft computing tools exploit the reasoning, intuition, consciousness, wisdom, and adaptability to changing environments possessed by human beings for developing computing paradigms like FL, ANNs, GAs, and SVMs. The integration of these techniques into the computing environment has given impetus to the development of smarter and wiser machines possessing logical and intuitive information processing capabilities equivalent to human beings. These techniques, whether complementing each other or working on their own, are able to model complex or unknown relationships which are either nonlinear or noisy. Soft computing techniques have a self-adapting characteristic, paving the way for the development of automated design systems. A synergistic partnership exploiting the strengths of these individual techniques can be harnessed for developing hybrid computing tools.

Among the forerunners in the field of soft computing is the Artificial Neural Network (ANN). Inspired by the functioning of a human brain, they have immense potential in modeling functional relationships which are either too complex or unknown in nature. The ANN model is a system of interconnected computational neurons arranged in an organized fashion to carry out the extensive computing necessary to perform the mathematical mapping. Unlike conventional methods of computing which are based on predefined rules, ANNs work on vague functional relationships through a process of learning from experience and examples.

Fuzzy Logic (FL) was conceptualized by Zadeh in the year 1965. It was inspired by how human beings make decisions when dealing with knowledge that is inexact, imprecise, and vague in nature; FL in a way emulates human expertise in solving a particular problem. Genetic Algorithms (GAs) are stochastic search and optimization tools which aim to find the optimal solution to a problem which has many sub-optimal solutions. They require little information about the problem to be solved and can effectively work with complex constraints and discrete variables. GAs working on the operators of natural evolution viz., reproduction, crossover, and mutation were inspired by Darwin's principle of 'survival of the fittest' and are able to find the global optimal solution to a particular problem.

In recent years, a novel machine learning technique called Support Vector Machine (SVM), proposed by Vapnik and based on statistical learning theory, has been successfully applied to pattern classification and regression estimation problems. Initially, SVMs were developed for solving classification problems and later were expanded by Vapnik to regression problems. Unlike ANNs, which are based on the Empirical Risk Minimization (ERM) induction principle, SVM implements the Structural Risk Minimization (SRM) principle. It is well known that ANN has the problem of overfitting and that the solution may get trapped in local minima, whereas the SRM principle aims to minimize an upper bound on the generalization error; and thus SVM will have higher prediction capabilities on unseen data sets. Further, SVM formulation will lead to the solution of a quadratic optimization problem with linear inequality constraints and the problem will have a global optimal solution. Combined with high generalization ability, SVM becomes a very attractive method.

This book is oriented to the approach of data-driven modeling to both spatial and temporal field data. A small component of the book related to fundamentals is devoted to machine learning algorithms. The main focus is on various models that have been utilized by the authors in various domains of Civil Engineering. This book is targeted to graduate and doctoral students of Civil Engineering, as well as Earth sciences departments and researchers interested in machine learning methods and applications.

The objective of machine learning (ML) is to develop algorithms which allow computers to learn. This learning is based on experience gained and abilities to generalize previous scenarios into new conditions. Learning abilities are important for human intelligence. In artificial intelligence, the main challenge is to equip the machine with this ability, either by implementing a set of algorithms or by a stand-alone robot.

In the early days of ML, the idea was to use it to make machines adaptable, interactive, and able to learn from experience. ML continues to bring many challenging problems to fundamental research. Speech recognition, computer vision, and feature extraction are the new scientific branches where ML is essential. Computational learning, which studies the properties of learning from the experimental/empirical/field data, is also becoming an important research field.

The importance of ML may be attributed to scientific research changes over the past few decades. The gathering of relevant data in the world around us has drastically improved with the support of recent technological advances, and Civil Engineering is a field which immensely benefits from these advances. Various sensors are put into systems to measure countless parameters, and they can be organized in wireless networks to provide huge volumes of information in real time.

The storage of data is a technical and engineering problem, whereas the understanding of the underlying phenomena is a scientific challenge. Recently, scientific research has been inclined towards a data-driven approach and this data-driven modeling has to respond to the problem accordingly, with the help of ML.

1.2 Learning from Data

At the beginning of learning, developing a system or an algorithm which can learn and generalize from the data is the formulation of an appropriate mathematical framework. In the majority of cases, an observation can be presented as a pair of entities—one is input space and the other is output space. Hence, empirical knowledge can be formulated as a set of these input–output pairs. Both input and output data can be encoded as multidimensional vectors.

From the observation and collected data, it can be assumed that some kind of underlying phenomena links inputs to outputs. This dependence is assumed as D, such that D maps input x to y. A deterministic mapping is a natural way to link vector spaces, although it is not realistic to restrict real-world processes which generate the data to be ideally deterministic. As there are many factors influencing data and measurement processes in the real world, the settings become stochastic. Hence, the role of probabilistic distribution P(x,y) is inevitable for representation of the generated data in order to achieve an acceptable description of the process. Usually, the explicit form of this type of distribution is unknown. However, some kind of inference from the available data set (x,y) generated by P(x,y) is to make sure that the data set is consistent and properly represented to provide reliable knowledge about P(x,y). Hence, it can be assumed that (x,y) are independent and identically distributed data sampled from the same population.

Machine learning developed algorithms are able to predict the outputs for previously unknown inputs without making restrictive assumptions about P(x,y). Although, whereas some ideas are purely algorithmic and distribution independent, it is essential to have an empirical data set that is representative of the underlying processes. This means the new samples are from the very same distribution/population for prediction.

1.3 Research in Machine Learning: Recent Progress

The field of engineering is a creative one. The problems encountered in this field are generally unstructured and imprecise, influenced by intuitions and past experiences of a designer. The conventional methods of computing that rely on analytical or empirical relations become time-consuming and labor-intensive compared with real-life problems. To study, model, and analyze such problems, approximate computer-based soft computing techniques inspired by the reasoning, intuition, consciousness, and wisdom possessed by human beings are employed. In contrast to conventional computing techniques, which rely on exact solutions, soft computing aims to exploit a given tolerance of imprecision, yielding an approximate solution to a problem in a short time. Soft computing, a multidisciplinary field, uses a variety of statistical, probabilistic, and optimization tools which complement each other to produce its three main branches viz., Neural Networks, Genetic Algorithms, and Fuzzy Logic. This book presents the applications of four major soft computing techniques viz., Artificial Neural Networks, Fuzzy Logic, Genetic Algorithms, and Support Vector Machines in the field of Civil Engineering, which to some extent has replaced the time-consuming conventional techniques of computing with intelligent and time–saving computing tools.

Civil Engineering is a diversified field ranging from water-resources to design and analysis of structures. The nature of the problems encountered in this field is of a complex nature and mostly requiring human intervention in the form of past experience and intuition. The heuristic nature of problems poses challenges to a civil engineer thereby making him resort to time and resource-saving computational tools. Soft computing techniques viz., Neural Networks, Fuzzy Logic, and Genetic Algorithms either working independently or complementing each other support the engineering activities by harnessing the cognitive behavior of the human mind to arrive at cost-effective solutions.

1.4 Artificial Neural Networks

Artificial Neural Networks are computational models designed to mimic the learning abilities of a human brain. Haykin described a neural network as a massively parallel distributed processor made up of simple processing units, which has a natural propensity for storing experiential knowledge and making it available for use. An ANN can be regarded as an engineering counterpart of a biological neuron. The interconnected processing units are called artificial neurons and replicate the functioning of biological neurons. Individually, the neurons perform trivial functions, but collectively, in the form of a network, they are capable of solving complicated problems. ANNs rely on past knowledge and when presented with input–output data pairs they construct a functional relationship through a process of learning. The learning ability of neural networks is attributed to the adjustment in the intensity of inter-neuron connection or the synaptic weight value. Adaptability to changing input–output data, nonlinear function mapping, and the ability to capture unknown relationships makes the ANN a versatile tool for modeling real-world problems. Chapter 8 elaborates on neural network applications in the fields of Civil Engineering, such as water resources, geotechnical, concrete, environmental, structural, transportation, coastal, etc.

1.5 Fuzzy Logic (FL)

Fuzzy Logic is similar to the human being's feeling and inference processes. Unlike classical control strategy, which is a point-to-point control, Fuzzy

Logic control is a range-to-point or range-to-range control. The output of a fuzzy controller is derived from fuzzifications of both inputs and outputs using the associated membership functions. A crisp input will be converted to the different members of the associated membership functions based on its value. From this point of view, the output of a Fuzzy Logic controller is based on its memberships of the different membership functions, which can be considered as a range of inputs. Fuzzy ideas and Fuzzy Logic are so often utilized in our routine life that nobody even pays attention to them. For instance, to answer some questions in certain surveys, most of the time one could answer with 'Not Very Satisfied' or 'Quite Satisfied,' which are also fuzzy or ambiguous answers. Exactly to what degree is one satisfied or dissatisfied with some service or product for those surveys? These vague answers can only be created and implemented by human beings, but not machines. Is it possible for a computer to answer those survey questions directly as a human being did? It is absolutely impossible. Computers can only understand either '0' or '1,' and 'HIGH' or 'LOW.' Those data are called crisp or classic data and can be processed by all machines.

The idea of Fuzzy Logic was invented by Professor L. A. Zadeh of the University of California at Berkeley in 1965. This invention was not well recognized until Dr. E. H. Mamdani, a professor at London University, applied the Fuzzy Logic in a practical application to control an automatic steam engine in 1974, which is almost ten years after the fuzzy theory was invented.

Fuzzy Logic is considered a logical system that provides a model for human reasoning modes that are approximate rather than exact. The Fuzzy Logic system can be used to design intelligent systems on the basis of knowledge expressed in human language. There is practically no area of human activity left untouched by intelligent systems as these systems permit the processing of both symbolic and numerical information. The systems designed and developed based on Fuzzy Logic methods have been proved to be more efficient than those based on conventional approaches.

Fuzzy Logic has found multiple successful applications, mainly in control theory. Fuzzy rule-based systems can be built by interviewing human experts or by processing historical data and thus forming a data-driven model. Fuzzy Logic applications are found in almost every domain of Civil Engineering such as water resources, geotechnical, structures, environmental, construction, transportation, etc.

1.6 Genetic Algorithms

Problems faced in the field of engineering are multi-faceted, primarily involving evaluation of the optimal solution to a problem governed by a number of numerical constraints of complex nature. The level of difficulty to choose the best among the available alternatives has compelled many engineers to detach from conventional techniques of analysis and move towards nature-inspired computational tools which have features like reproduction and self- adaptation to a changing environment. Genetic Algorithms are a class of stochastic optimization techniques which work on the principle of evolution. According to Koza (1992), 'The Genetic Algorithm is a highly parallel mathematical algorithm that transforms a set (population of individual mathematical objectives typically fixed-length character strings patterned after chromosome strings), each with an associated fitness value, into a new population (i.e., the next generation) using operations patterned after the Darwinian principle of reproduction and survival of the fittest and after naturally occurring genetic operations (notably sexual recombination).' The process of optimization lies at the root of engineering, since the classical function of the engineer is to design new, better, more efficient, and less expensive systems, as well as to devise plans and procedures for the improved operation of existing systems. The application of GA in the field of Civil Engineering primarily deals with optimization problems governed by a mixed nature of variables. This is in contrast to conventional optimization techniques which are based on the steepest gradient descent approach and applicable to a continuous nature of variables. This paradigm shift in the methodology to solve multi-objective problems influenced by sub-optimal solutions has created interest in the field of directed search algorithms which not only aim at feasible design but also cater to the requirements of the design objective. Similar to ANN and FL, the applications of GA in the field of Civil Engineering are widely observed in areas such as structural engineering, concrete mix design, geotechnical engineering, water resources, transportation engineering, environmental engineering, etc.

1.7 Support Vector Machine (SVM)

SVM is a relatively new statistical learning technique. Due to its strong theoretical statistical framework, SVM has proved to be much more robust in several fields, especially for noise mixed data, than the local model which utilizes traditional chaotic techniques. SVM has brought forth heavy expectations in the past few years as they have been successful when applied in classification problems, regression, and forecasting; as they include aspects and techniques from machine learning, statistics, mathematical analysis, and convex optimization. Apart from possessing a strong adaptability, global optimization, and a good generalization performance, the SVMs are also suitable for classification of small samples of data. Globally, the application of these techniques in the field of hydrology has come a long way since the first articles began appearing in conferences in the early 2000s.

Since training an SVM classifier requires solving a large quadratic programming problem with linear inequality constraints, a popular algorithm called Sequential Minimal Optimization (SMO) was proposed by Platt. In this algorithm the large quadratic programming problem is broken into a series of the smallest possible quadratic programming problems. As an improvement, the SMO algorithm suggested a method of avoiding the use of threshold value at the time of checking the Karush–Kuhn–Tucker (KKT) conditions. Following the idea of decomposing the large problem into a series of smaller tasks, a new strategy has been developed in the implementation of the SVM learning algorithm. In this algorithm the original optimization problem is split into active and inactive parts. This algorithm differs from SMO in the working set selection strategies. SVMlight solves for both the pattern classification and regression problems. LIBSVM is another software used for solving support vector classification and regression problems. Also, it supports multi-class SVM. It implements an SMO type of algorithm. SVMs have been successfully applied to a large number of class of problems of importance in Civil Engineering domains such as water resources, structural engineering, concrete mix design, geotechnical engineering, etc.

1.8 Hybrid Approach (HA)

The hybrid approach combines two or more of the aforementioned approaches above to develop a hybrid model. It is a multiuse and powerful tool for modeling complex processes and characterizing uncertainty in quality evaluation. The following hybrid combinations are popular in the domain of Civil Engineering: Fuzzy-ANN, Fuzzy-GA, GA-ANN, GA-SVM, Fuzzy-SVM, etc.

Bibliography

Abrahart, R.J. and See, L.M. 2007. Neural network modeling of non-linear hydrological relationships, *Hydrology and Earth System Sciences*, 11, 1563–1579.

Adeli, H. and Park, H.S. 1995. A neural dynamics model for structural optimization-theory, *Computers and Structures*, 57(3), 383–390.

Adib, A. and Jahanbakhshan, H. 2013. Stochastic approach to determination of suspended sediment concentration in tidal rivers by artificial neural networks and genetic algorithms, *Canadian Journal of Civil Engineering*, 40(4), 299–312.

Ahmadkhanlou, F. and Adeli, H. 2005. Optimum cost design of reinforced concrete slabs using neural dynamics model, *Engineering Applications of Artificial Intelligence*, 18, 65–72.

Alqedra, M., Arafa, M., and Ismail, M. 2011. Optimum cost of prestressed and rein-
    forced concrete beams using genetic algorithms, *Journal of Artificial Intelligence*,
    4(1), 76–88.
Amirjanov, A. and Sobolev, K. 2005. Optimal proportioning of concrete aggregates
    using a self adaptive genetic algorithm, *Computers and Concrete*, 2(5), 1–11.
Arslan, H.M. 2010. An evaluation of effective design parameters on the earthquake perfor-
    mance of RC buildings using neural networks, *Engineering Structures*, 32, 1888–1898.
Aydin, Z. and Ayvaz, Y. 2013. Overall cost optimization of prestressed concrete
    bridges using genetic algorithm, *KSCE Journal of Civil Engineering*, 17(4), 769–776.
Bhattacharjya, R. 2004. Optimal design of unit hydrographs using probability distri-
    butions and genetic algorithms, *Sadhana*, 29(5), 499–508.
Camp, C., Pezeshk, S., and Hansson, H. 2003. Flexural design of reinforced concrete
    frames using a genetic algorithm, *Journal of Structural Engineering*, 129(1), 105–115.
Chau, K.W. and Albermani, F. 2002. Genetic algorithms for design of liquid retaining
    structures, *Lecture Notes in Artificial Intelligence*, 2358, 119–128.
Chen, Y.W., Cheng, L.C., Huang, C.W., and Chu, H.J. 2013. Applying genetic algo-
    rithms and neural networks to the conjunctive use of surface and subsurface
    water, *Water Resources Management*, 2013.
Cheng, C., Wu, X., and Chau, K.W. 2005. Multiple criteria rainfall-runoff model cali-
    bration using a parallel genetic algorithm in a cluster of computers, *Hydrological
    Sciences Journal*, 50(6), 1069–1087.
Cho, H.C., Fadali, M.S., Saiidi, S.M., and Soon, L.K. 2005. Neural network active con-
    trol of structures with earthquake excitation, *International Journal of Control,
    Automation and Systems*, 3(2), 202–210.
Cladera, A. and Mari, A.R. 2004. Shear design procedure for reinforced normal and
    high-strength concrete beams using artificial neural networks. Part II: Beams
    with stirrups, *Engineering Structures*, 26, 927–936.
Cortes, C. and Vapnik, V. 1995. Support-vector networks, *Machine Learning*, 20(3),
    273–297.
Cui, L. and Sheng, D. 2005. Genetic algorithms in probabilistic finite element analysis
    of geotechnical problems, *Computers and Geotechnics*, 32, 555–563.
Das, S., Manna, B., and Baidya, D.K. 2011. Prediction of dynamic soil-pile interaction
    under coupled vibration using artificial neural network approach, *Geo Frontiers*,
    1–10.
Deb, K. 1991. Optimal design of a welded beam via genetic algorithms, *AIAA Journal*,
    29, 2013–2015.
Drucker, H., Wu, S., and Vapnik, V.N. 1999. Support vector machines for spam catego-
    rization, *Neural Networks, IEEE Transactions*, 10(5), 1048–1054.
Flood, I. and Kartam, N. 1994. Neural networks in civil engineering. I: Principles and
    understanding, *Journal of Computing in Civil Engineering*, 8(2), 131–148.
Fu, G. and Kapelan, Z. 2011. Embedding neural networks in multiobjective genetic
    algorithms for water distribution system design, *Water Distribution Systems
    Analysis*, 2010, 888–898.
Fu, K., Zhai, Y., and Zhou, S. 2005. Optimum design of welded steel plate girder
    bridges using a genetic algorithm with elitism, *Journal of Bridge Engineering*,
    10(3), 291–301.
Ghaboussi, J. and Elnashai, A.S. 2007. Development of neural network based hyster-
    etic models for steel beam-column connections through self learning simula-
    tion, *Journal of Earthquake Engineering*, 11, 453–467.

Golabi, M., Radmanesh, F., Akhondali, A., and Kashefipoor, M. 2013. Prediction of seasonal precipitation using artificial neural networks (Case study: Selected stations of (Iran) Khozestan Province), *Journal of Basic and Applied Scientific Research*, 3(1), 589–595.

Goldberg, D.E. and Samtani, M.P. 1986. Engineering optimization via genetic algorithm, *Proceedings 9th Conference on Electronic Computation, ASCE*, pp. 471–482.

Guang, N.H. and Zong, W.J. 2000. Prediction of compressive strength of concrete by neural networks, *Cement and Concrete Research*, 30, 1245–1250.

Guerra, A. and Kiousis, D.P. 2006. Design optimization of reinforced concrete structures, *Computers and Concrete*, 3(5), 313–334.

Gupta, R., Kewalramani, M., and Goel, A. 2006. Prediction of concrete strength using neural-expert system, *Journal of Materials in Civil Engineering*, 18(3), 462–466.

Hadi, M.N.S. 2003. Neural network applications in concrete structures, *Computers and Structures*, 81(6), 373–381.

Hajela, P. and Berke, L. 1991. Neurobiological computational models in structural analysis and design, *Computers and Structures*, 41(4), 657–667.

Haykin, S. 2009. *Neural Networks: a Comprehensive Foundation*, 8th ed., Pearson Prentice Hall, New Delhi, India.

Hejazi, F., Toloue, I. Jaafar, M.S., and Noorzaei, J. 2013. Optimization of earthquake energy dissipation system by genetic algorithms, *Computer Aided Civil and Infrastructure Engineering*, 28(10), 796–810.

Heng, S. and Suetsugi, T. 2013. Using artificial neural networks to estimate sediment in ungauged catchment of the Tonle Sap River Basin Cambodia, *Journal of Water Resources and Protection*, 5, 111–123.

Jang, J.S.R., Sun C.T., and Mizutan E. 1996. *Neuro-fuzzy and Soft Computing: A Computational Approach to Learning and Machine Intelligence*, Prentice Hall, chapters 17–21, pp. 453–567.

Jain, A., Bhattacharjya, R., and Sanaga, S. 2004. Optimal design of composite channels using genetic algorithms, *Journal of Irrigation and Drainage*, 130(4), 286–295.

Jain, A. and Srinivasulu, S. 2004. Development of effective and efficient rainfall-runoff models using integration of deterministic, real coded genetic algorithms and artificial neural networks, *Water Resources Research*, 40(4).

Jakubek, M. 2012. Neural network prediction of load capacity for eccentrically loaded reinforced concrete columns, *Computer Assisted Methods in Engineering and Science*, 19, 339–349.

Jayaram, M.A., Nataraja, M.C., and Ravikumar, C.N. 2009. Elitist genetic algorithm models: optimization of high performance concrete mixes, *Materials and Manufacturing Processes*, 24(2), 225–229.

Jenkins, W. 1992. Plane frame optimum design environment based on genetic algorithm, *Journal of Structural Engineering*, 118(11), 3103–3112.

Jeong, D. and Kim, Y. 2005. Rainfall-runoff models using artificial neural networks for ensemble streamflow prediction, *Hydrological Processes*, 19, 3819–3835.

Karasekreter, N., Basciftci, F., and Fidan, U. 2013. A new suggestion for an irrigation schedule with an artificial neural network, *Journal of Experimental and Theoretical Artificial Intelligence*, 25(1), 93–104.

Karrey, F. and de Silva, C. 2004. *Soft Computing and Intelligent System Design*, Addison Welsley, chapter 7, pp. 337–361.

Kim, S., Choi, H.B., Shin, Y., Kin, G.H., et al. 2013. Optimizing the mixing proportions with neural networks based on genetic algorithms for recycled aggregate concrete, *Advances in Material Sciences and Engineering*, 2013.

Koumousis, V.K. and Arsenis, S.J. 1994. Genetic algorithms in a multi-criterion optimal detailing of reinforced concrete members, *Advances in Structural Optimization*, CIVIL-COMP Ltd, Edinburgh, Scotland, pp. 233–240.

Koza, J.R. 1992. *Genetic Programming: On the Programming of Computers by Means of Natural Selection*, MIT Press, Cambridge, MA.

Kumar, M., Raghuwanshi, N., Singh, R., Wallender, W., et al. 2002. Estimating evapotranspiration using artificial neural network, *Journal of Irrigation and Drainage Engineering*, 128(4), 224–233.

Lee, S.C. and Han, S.W. 2002. Neural network based models for generating artificial earthquakes and response spectra, *Computers and Structures*, 80(20–21), 1627–1638.

Liu, M., Qie, Z., Wu, X., Dong, W., et al. 2008. Model optimization of load-bearing capacity of macadam pile composite foundation based on genetic algorithms, *IEEE Control and Decision Conference, Yantai*, pp. 3903–3907.

Liu, X., Cheng, G., Wang, B., and Lin, S. 2012. Optimum design of pile foundation by automatic grouping genetic algorithms, *ISRN Civil Engineering*, 2012.

Manouchehrian, A., Gholamnejad and Sharifzadeh, M. 2013. Development of a model for analysis of slope stability of circular mode failure using genetic algorithms, *Environmental and Earth Sciences* 71(3), 1267–1277.

Mitchell, T. 1997. *Machine Learning*, McGraw Hill Education, India.

Mukherjee, A. and Despande, J.M. 1995. Modeling initial design process using artificial neural networks, *Journal of Computing in Civil Engineering*, 9(3), 194–200.

Naeim, F., Alimoradi, A., and Pezeshk, S. 2004. Selection and scaling of ground motion time histories for structural design using genetic algorithms, *Earthquake Spectra*, 20(2), 413–426.

Nagy, H., Watanabe, K., and Hirano, M. 2002. Prediction of sediment load concentration in rivers using artificial neural network model. *Journal Hydraulic Engineering*, 128(6), 588–595.

Nath, U.K. and Hazarika, P.J. 2011. Study of pile cap lateral resistance using artificial neural networks, *International Journal of Computer Applications*, 21(1), 20–25.

Nikoo, M., Zarfam, P., and Nokoo, M. 2012. Determining the displacement in concrete reinforcement building using evolutionary artificial neural networks, *World Applied Sciences Journal*, 16(2), 1699–1708.

Nixon, J.B, Dandy, G.C., and Simpson, A.R. 2001. A genetic algorithm for optimizing off-farm irrigation scheduling, *Journal of Hydroinformatics*, 3(1), 11–22.

Noguchi, T. and Maruyama, I. 2003. Performance based design system for concrete with multi-optimizing genetic algorithm, *Proceedings of 11th International Congress on Chemistry of Cement*, Durban, South Africa.

Oreta, A.W.C. and Kawashima, K. 2003. Neural network modeling of confined compressive strength and strain of circular concrete columns, *Journal of Structural Engineering*, 129(4), 554–561.

Ozturan, M., Birgul, K., and Ozturan, T. 2008. Comparison of concrete strength prediction techniques with artificial neural network approach, *Building Research Journal*, 56(1), 23–36.

Rafiq, M., Bugmann, G., and Easterbrook, D. 2001. Neural network design for engineering applications, *Journal of Computers and Structures*, 79(17), 1541–1552.

Raghavendra, N.S., Deka, P.C. 2014. Support vector machine applications in the field of hydrology: a review. *Applied Soft Computing* 19, 372–386. doi:10.1016/j.asoc.2014.02.002.

Rahman, M.M. and Jumaat, M.Z. 2012. Cost minimum proportioning of non-slump concrete mix using genetic algorithms, *Advanced Materials Research*, 468–471, 50–54.

Rai, R.K. and Mathur, B.S. 2008. Event based sediment yield modeling using artificial neural networks, *Water Resources Management*, 22(4), 423–441.

Rajasekharan, S. and Vijaylakshmi Pai, G.A. 2005. Neural Network, Fuzzy Logic and Genetic Algorithm-synthesis and Applications, Prentice Hall, chapters 10–15, pp.297–435.

Rajeev, S. and Krishnamoorthy, C. 1992. Discrete optimization of structures using genetic algorithms, *Journal of Structural Engineering*, 118(5), 1233–1250.

Rao, H.S. and Babu, B.R. 2007. Hybrid neural network model for the design of beam subjected to bending and shear, *Sadhana*, 32(5), 577–586.

Ravindran, A., Ragsdell, K.M. and Reklaitis, G.V. 2006. *Engineering Optimization: Methods and Applications*, John Wiley & Sons Inc, Hoboken, NJ.

Robles, C.M.A. and Becerril, H.R.A. 2012. Seismic alert system based on artificial neural networks, *World Academy of Science, Engineering and Technology*, 66, 813–818.

Rogers, J.L. 1994. Simulating structural analysis with neural network, *Journal of Computing in Civil Engineering*, 8(2), 252–265.

Sambridge, M. and Gallagher, K. 1993. Earthquake hypocenter location using genetic algorithms, *Bulletin of the Seismological Society of America*, 83(5), 1467–1491.

Sanad, A. and Saka, M. 2001. Prediction of ultimate shear strength of reinforced-concrete deep beams using neural networks, *Journal of Structural Engineering*, 127(7), 818–828.

Sarma, K. and Adeli, H. 1998. Cost optimization of concrete structures. *Journal of Structural Engineering*, 124(5), 570–578.

Senouci, B.A., and Al-Ansari, S.M. 2009. Cost optimization of composite beams using genetic algorithms, *Advances in Engineering Software*, 40, 1112–1118.

Shahin, M.A., Maier, H.R., and Jaksa, M.B. 2002. Predicting settlement of shallow foundations using neural networks, *Journal of Geotechnical and Geoenvironmental Engineering*, 128(9), 785–793.

Tabari, H., Marofi, S., and Sabziparvar, A. 2010. Estimation of daily pan evaporation using artificial neural networks and multivariate non-linear regression, *Irrigation Science*, 28(5), 399–406.

Tokar, S.A. and Johnson, P.A. 1999. Rainfall-runoff modeling using artificial neural networks, *Journal of Hydrologic Engineering*, 4(3), 232–239.

Vanluchene, R.D. and Sun, R. 1990. Neural networks in structural engineering, *Computer-Aided Civil and Infrastructure Engineering*, 5(3), 207–215.

Vapnik, V.N. (1995) *The Nature of Statistical Learning Theory*, Springer Verlag, New York.

Wilby, R.L. Abrahart, R.J., and Dawson, C.W. 2003. Detection of conceptual model rainfall-runoff processes inside an artificial neural network, *Hydrological Sciences Journal*, 48(2), 163–181.

Wongprasert, N. and Symans, M. 2004. Application of a genetic algorithm for optimal damper distribution within the nonlinear seismic benchmark building, *Journal of Engineering Mechanics*, 130, 401–406.

Xie, J., Qiu, J., Wei, L. and Wang, J.W. 2011. The application of neural network in earthquake prediction in East China, *Advances in Intelligent and Soft Computing*, 106, 79–84.

Xie, X., Yan, D., and Zheng, Y. 2011. Optimization design of high performance concrete based on genetic algorithm toolbox of Matlab, *Advanced Materials Research*, Vol. 250–253, 2672–2677.

Yeh, I. (2006). Exploring concrete slump model using artificial neural networks, *Journal of Computing in Civil Engineering*, 20(3), 217–221.

Yilmaz, S. 2011. Ground motion predictive modeling based on genetic algorithms, *Natural Hazards and Earth System Sciences*, 11, 2781–2789.

Yu, T., Frenandez, J., and Rundle, J.B.1998. Inverting the parameters of an earthquake rupture fault with a genetic algorithm, *Computers and Geosciences*, 24(2), 173–182.

Zadeh, L.H. 1965. Fuzzy sets, *Information and Control*, 8, 338–353.

Zhang, F.X., Wai, O.W.H., and Jiang, Y.W. 2010. Prediction of sediment transportation in Deep Bay (Hong Kong) using genetic algorithms, *Journal of Hydrodynamics Series B*, 22(5), Supplement 1, 599–604.

Zhang, Q. and Wang, C. 2008. Using genetic algorithm to optimize artificial neural networks: a case study on Earthquake prediction, *IEEE Second International Conference on Genetic and Evolutionary Computing*, Hubei, pp. 128–131.

2

Artificial Neural Networks

2.1 Introduction to Fundamental Concepts and Terminologies

The architecture of the human brain is highly intricate, nonlinear, and consists of parallel processing systems. It is capable of performing a given task many times faster than any computers in existence. Therefore, modeling the functions of the human brain is highly complex and challenging. The research on Artificial Neural Networks (ANNs) has been inspired by referring to the way the human brain works. ANNs are the abstract of the human brain's systems. A neural computing system is a parallel distribution process. The complex computation in an ANN can be operated by introducing individual computing units where each individual unit performs an elementary computation. The attributes of neural networks are architecture and functional properties or neurodynamics. The architecture refers to the network and the number of artificial neurons present in the network and their interconnectivity. As the name implies, a neural network is a structure with many nodes which are connected through directional links. They are the interconnected neurons (or processing units) with known characteristics, such as inputs, synaptic strengths, activation, outputs, and bias. The neurodynamics of a neural network refers to the properties of each unit present in the network. These properties include how it learns, recalls, associates, and continuously compares the new information with existing knowledge fed to the network.

The concepts of ANNs were first applied by Rosenblatt for *single-layer perceptrons* in order to pattern classification learning in the late 1950s. ANNs are capable of performing supervised and/or unsupervised learning. Human thinking is a parallel activity with numerous neurons connected together. The main characteristics of human thinking are imprecision, fuzziness, and adaptability.

Terminologies of ANNs include the following:

Processing Unit: An ANN consists of interconnecting processing units with a summing part followed by an output unit. The summing part receives input values, weights of each input value, and weighted sums which are known as activated values. The output part then produces a signal with these activated values.

Interconnections: In ANN, several processing units are interconnected in order to accomplish some pattern recognition task. The input to a processing unit may either come from the output of other processing units and/or from some other external sources. The output received due to the processing of another unit is strongly influenced by the strength and the weight associated with that connection link.

Operations: During the period of operations, each ANN unit will receive input from other connected units and/or from some external sources. A sum of inputs is computed at a given instant of time. The activation value determines the actual output from the output function unit (i.e., the output state of the units). The output values and other external inputs, in turn, determine the activation and output states of other units. Activation dynamics determine the activation values of all the units (i.e., the activation state of the network as a function of time). The activation dynamics also determine the dynamics of the output state of the network. The set of all output states defines the output state space of the network.

Update: In the implementation, there are several options available for both activation and synaptic dynamics. In particular, the updating of the output states of all the units could be performed synchronously. In this case the activation values of all the units are computed at the same time, assuming a given output state throughout.

2.2 Evolution of Neural Networks

The historic developments in the field of neural networks are outlined in Table 2.1.

2.3 Models of ANN

The development of mathematical models are a prerequisite for the creation of ANN, which are capable to best derive the system functions. The neuron is the basic cellular unit of the nervous system in the human brain. Each neuron consists of an output fiber called an axon and a button-like terminal called the synapse region. The axons are split up and again connected to many dendrites to act as an input path for other neurons. Each neuron receives and combines the various signals from other neurons connected through dendrites. A simple model of a neuron is the fundamental processing element of a neuron network. The use of computers will simulate the model quickly. Changes can be applied to the model to improve the performance or to simplify the task, etc. A fair knowledge of the system is essential for the model creation.

TABLE 2.1

The Evolution of Artificial Neural Networks

Author and Year	Contributions
McCulloch and Pitts, 1943	Model of the computing element called McCulloch–Pitts neuron model, where the weights are fixed; hence the model could not learn from the examples.
Hebb, 1949	Hebb's law, a fundamental learning rule in neural networks by adjusting a connection weight.
Minsky, 1954	Development of learning machine through which connection strength could be adapted automatically.
Rosenblatt, 1958	Proposed perceptron model which has weight adjustable by perceptron learning law.
Widrow and Hoff, 1960	Proposed the ADALINE model and LMS algorithm to adjust the weights. The algorithm was successfully used for adaptive signal processing systems.
Hopfield, 1982 and Hopfield, 1984	Energy analysis of feedback neural network systems.
Kohenen, 1982	Developed self-organizing feature maps.
Ackley et al., 1985	Proposed a feedback neural network with stochastic neurons called the Boltzmann machine.
Rumelhart et al., 1986	Proposed the possibility of adjusting the weights in feed-forward neural networks.
Linsker, 1988	Self-organization based on information preservation.
Kosko, 1988	Fuzzy logic in ANN.
Poggio and Girosi, 1990	Radial basis functions and regularization theory.

2.4 McCulloch–Pitts Model

The McCulloch–Pitts Model (MP) is the simplest ANN and is also known as a threshold logic unit. In an MP model, the activation (x) is given by a weighted sum of its M input values (a_i) and bias term (Θ), where (s) is the output signal which is a nonlinear function $f(x)$ of the activation value as explained in Figure 2.1 below.

In an MP model, the weights are fixed, therefore the model does not have the capability of learning, and it will produce only binary output.

2.5 Hebb Network

Donal Hebb (Hebb, 1949) proposed a neural learning pattern for updating synaptic weights for the associative memories, which is now known as the Hebbian learning rule. As per this rule, the information can be stored in synaptic weights and assume the learning techniques that had an impact

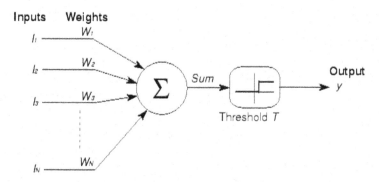

FIGURE 2.1
McCulloch–Pitts model.

on the development in the field of neural learning and associative memories. The Hebbian rule is the most popular rule for adjusting the weights of binary neural networks. The original statement of Hebb (1949) is quoted as 'When an axon in a cell A near enough to excite a cell B, and repeatedly or persistently takes part in firing it, some growth process or metabolic change takes place in one or both cells such that A's efficiency as one of the cells firing B is increased.' Here the weights between the neurons whose activities are positively correlated are increased.

2.6 Summary

In this section, the basics of an ANN and its growth have been discussed. The comparison between a biological neuron and an artificial neuron has been included. The basic terminologies of ANN are discussed. A brief description of the McCulloch–Pitts neuron model is included. Also, details are provided for the Hebb network.

2.7 Supervised Learning Network

Supervised neural networks are the mainstream of neural network development. The differentiable characteristic of a supervised neural network lies in the inclusion of a teacher in the learning process. For the learning process, the network needs training data with examples consisting of a various number of input–output combinations. The desired output in the training data sets will then act as a teacher for the network's learning process. An error signal is then defined by considering the expected output and the system output. The learning process will continue until the error is close to zero. The sum of all the errors is considered a network performance measure. The supervised network's architecture depends on the complexity and nature of the data handled. The block diagram of supervised learning for all neural networks is shown in Figure 2.2.

2.7.1 Perceptron Network

The perceptron model is the simple neural network developed by Frank Rosenblatt in 1962. It has more than one input connected to a node summing the linear combination of the inputs connected to the node. The resulting sum goes through a hard limiter which produces an output of +1 if the input of the hard limiter is positive. Similarly, it produces an output of –1 if the input is negative. It was first developed to classify a set of external inputs into two classes of C1 or C2 with an output +1 signifying C1 or C2. Despite its early success, it was proven by Minsky and Papert that the single-layer perceptron network is unable to classify linear non-separable problems.

2.7.2 Adaptive Linear Neuron

Adaptive Linear Neuron or Adaptive Linear Element (ADALINE) is a single-layer neural network. It was developed by Professor Bernard Widrow and Ted Hoff in 1960 and is based on the MP neuron with weights, bias, and summation functions. In the learning phase, the weights are adjusted according to the weighted sum of the inputs. In general, ADALINE can be trained by using the delta rule, which is also known as the least mean square (LMS) or the Widrow–Hoff rule. In ADALINE there will be only one input unit.

$$\text{net} = b + \sum_i xi\,wi$$

if net is being used for the pattern classification in which the desired output is either +1 or –1, then a threshold function is applied to the net input to obtain the activation.

$$y = f(\text{net}) = \begin{cases} 1 & \text{if } \text{net} \geq \theta \\ -1 & \text{if } \text{net} < \theta \end{cases}$$

$$b(\text{new}) = b(\text{old}) + a(t\text{-net})$$

$$wi(\text{new}) = w_i(\text{old}) + (t\text{-net})x_i$$

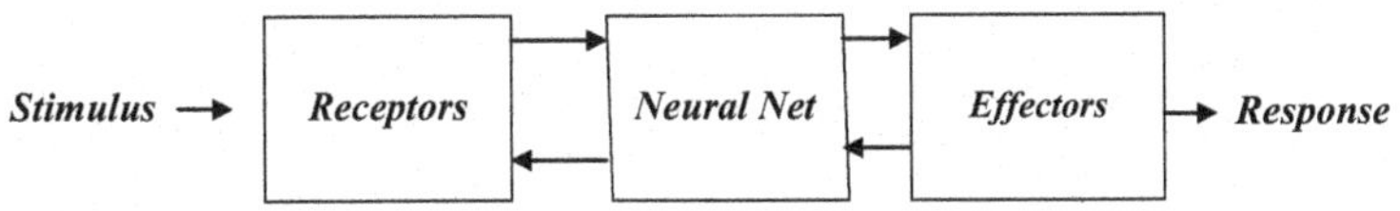

FIGURE 2.2
Block diagram representation of nervous system.

For a single neuron, the suitable value of the learning rate α is to be in the range of $0.1 \leq n^*\alpha \leq 1$, where n is the total number of the input units. The learning rule minimizes the mean squared error between the activation and the target value.

2.7.3 Back-Propagation Network

Many supervised learning procedures, especially multilayer neural networks have been applied successfully in different fields. The error backpropagation (BP) learning algorithm is one of the most popular. Although it was introduced in the 1970s, the importance of BP was truly recognized after the paper by David Rumelhart, Geoffrey Hinton, and Ronald Williams in 1986. BP works based on an iterative gradient which can be applied to minimize an error between the actual output vector of the network and the expected output vector. Adjustments to the weights are made in a sequence of steps. As the neural networks are employed for highly nonlinear applications, algorithms for supervised training are mainly based on the nonlinear optimization methods. For instance, the multilayer perceptron architecture is also sometimes known as a BP network. The term BP is also used to define the training of multilayer perceptrons using gradient descent applied to a sum of square error functions. Figure 2.3 shows feed-forward backpropagation (FFBP).

2.7.4 Radial Basis Function Network

Radial basis function (RBF) networks correspond to a particular class of function approximators which can be trained by using a set of samples. First

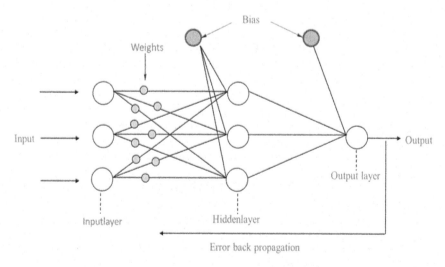

FIGURE 2.3
Feed-forward back propagation.

proposed by Broomhead and Lowe (1988) and Moody and Darken (1988), an RBF consists of three layers: input, hidden, and output. Compared to the multilayer perceptron (MLP), an RBF has only one hidden layer. The hidden units provide a set of functions that constitute an arbitrary basis for the input parameters. These units are also known as radial centers and are represented by vectors c1, c2, ...ch. The transformation from an input space to a hidden unit space is nonlinear while transformation from hidden units to output space is linear. The hidden layer produces a significant non-zero response only when the input falls within a small localized region of the input space. The output of the network is a linear combination of RBF of inputs and neuron parameters. They have many applications mainly in the field of function approximation, time series prediction, classification, and system control. An RBF performs classification by measuring the input's similarity to examples from the training set. Each RBF neuron stores a prototype which is one of the examples from the training set. To classify a new input, each neuron computes Euclidean distance between the input and its prototype. If the input is closer to A class prototype than B class prototype, then it classified as A. Figure 2.4 shows the structure of the radial basis function neural network (RBFNN).

2.7.5 Generalized Regression Neural Networks

Generalized regression neural networks (GRNNs) are single-pass associative memory feed-forward ANNs and use normalized Gaussian kernels in the hidden layer as activation functions. A GRNN is made of input, hidden, summation, division layer, and output layers as shown in Figure 2.5. When

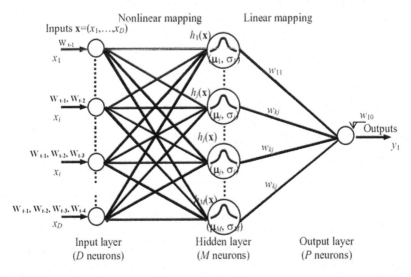

FIGURE 2.4
The structure of RBFNN.

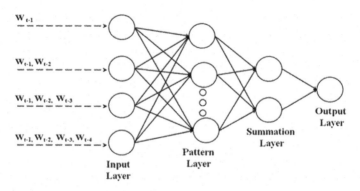

FIGURE 2.5
The structure of GRNN.

a GRNN is trained, it memorizes every unique pattern. This is the reason why it is a single-pass network and does not require any back-propagation algorithm. GRNN advantages include its quick training approach and its accuracy. On the other hand, one of the disadvantages of GRNN is the growth of the hidden layer size. However, this issue can be solved by implementing a special algorithm which reduces the growth of the hidden layer by storing only the most relevant patterns. Figure 2.5 shows the structure of a GRNN.

2.7.6 Summary

In this section, we discussed supervised learning networks, including perceptron, adaptive, backpropagation, and radial basis function. These networks are trained using delta learning rules. Back-propagation networks are the most widely used networks in real-time applications. The radial basis function network is mostly used in Gaussian activation function.

2.8 Unsupervised Learning Networks

2.8.1 Introduction

Unlike the supervised networks, in an unsupervised neural network the training of data sets is carried out without the supervision of a teacher. That means the learning process in unsupervised neural networks is carried out by a self-organizing pattern. During the training process, no external reference is used to adjust the weight of the network. The input vectors of similar patterns will try to form a cluster in a training process. When a new input is defined, then the neural network of unsupervised learning will produce an output based on the pattern to which it belongs. The correct output will not be available during the course of training. A typical unsupervised network consists of an input and a competitive layer. The neurons in the competitive layer will match with each other using some competitive rules during the learning period and produce the best output for the given input pattern. Self-organizing map (SOM) and Adaptive Resonance Theory (ART) are the most widely used neural networks in unsupervised learning algorithms. Unsupervised learning can be further classified into clustering and association problems. A clustering problem will create inherent grouping in the data based on its behavior. While in an association rule of learning, it will discover rules to describe the data, such as 'people who buy X also tend to buy Y.' k-means for clustering problems and a priori algorithms for association problems are some popular algorithms for unsupervised learning.

2.8.2 Kohonen Self-Organizing Feature Maps

Self-organizing feature maps (SOMs) are trained by using unsupervised learning to produce a low-dimensional, typically two-dimensional, discretized representation of input space of the training samples such as a map. Developed by Tuevo Kohonen in 1982, they differ from other neural networks as they apply competitive learning and use a neighborhood function to preserve the topological properties of the input space. The training of a SOM requires no target vector. A SOM learns to classify the training data without any external supervision. It is made up of input nodes and computational nodes. Each computational node is connected to input nodes to form a lattice and without any interconnections among the computational nodes. The number of input nodes are fixed based on the dimensions of the input vector.

2.8.3 Counter Propagation Network

A Counter Propagation Network (CPN) is an example of a hybrid network which is capable of combining two or more features of a network. First proposed by Hecht and Nielsen in 1986, it consists of a hidden layer (which works on the principle of the Kohonen network with unsupervised learning) and an output layer (which is a grossberg layer that is fully connected to the hidden layer. The output layer is trained by the Widrow–Hoff rule.

The output of the A subsection in the input layer is fanned out to the competitive middle layer. Each neuron in the output layer receives a signal corresponding to the input pattern's category along with one connection from the middle layer, while the B subsection of the input layer has zero input during actual operation of the network and is used only during the training process. The output layer will produce the pattern based on the output category in the middle layer. The output layer in the CPN will follow a supervised learning procedure with a direct connection with the input layer's B subsection and provides the correct output.

CPN follows a two-stage training procedure. First, the Kohonen layer is trained on the input patterns without making any changes for the output layer at this stage. Once the middle layer is trained to correctly categorize all the input patterns, the weights between the input and middle layers are kept fixed and the output layer is trained to produce correct output patterns by adjusting weights between the middle and output layers.

As a pattern associator, it is the advantage of a CPN to do the inverse mapping. But the training of a CPN has the same difficulty associated with the training of a Kohonen network. Moreover, CPNs are larger than BP networks (i.e., if a certain number of mappings are to be learned, the middle layer requires a large number of neurons).

2.8.4 Adaptive Resonance Theory Network

Adaptive Resonance Theory (ART) was proposed by Stephen Grossberg and Gail Carpenter and based on how the brain processes information.

Essentially, ART models incorporate new data by checking for similarity between this new data and data already learned (memory). If there is a close enough match, the new data is learned.

Otherwise, this new data is stored as a new memory. Some models of Adaptive Resonance Theory are:

- ART1—Discrete input.
- ART2—Continuous input.
- ARTMAP—Using two input vectors, transforms the unsupervised ART model into a supervised one.
- Various others—Fuzzy ART, Fuzzy ARTMAP (FARTMAP), etc.

2.8.5 Summary

In this section, various unsupervised learning networks have been discussed. Unsupervised learning networks are commonly used when the clustering of time units is performed. Here, information about the output is not known. The most important unsupervised learning network is the Kohonen self-organizing feature map, where clustering is performed over the training vectors and the network training is achieved. The counter propagation network is briefly discussed as a compression network. Similarly, Adaptive Resonance Theory network is also included.

2.9 Special Networks

2.9.1 Introduction

In this section, some specialized networks will be discussed: Gaussian machines, Cauchy machines, probabilistic neural networks, cascade correlation networks, cognition networks, cellular neural networks, and optical neural networks. Cauchy and Gaussian networks are the variation of fixed-weight optimization net. Probabilistic neural networks are designed using the probability theory to classify the input data. Cascade correlation networks are designed depending on the hierarchical arrangements of the hidden units. A cellular neural network (CNN), also known as a cellular nonlinear network, is an array of dynamical systems (cells) or coupled networks with local connections only.

2.9.2 Gaussian Machine

The Gaussian machine is based on the following three parameters: a slope parameter of a sigmoidal function, a time step, and temperature.

The noise which is found to obey a logistic, rather than a Gaussian distribution, produces a Gaussian machine. The output is set to one with probability. This does not bother about the unit's original state. When noise is added to the net input of a unit, then using probabilistic state transition gives a method for extending the Gaussian machine into a Cauchy machine.

2.9.3 Cauchy Machine

Cauchy machine can be called fast simulated annealing, and it is based on including more noise to the net input for increasing the likelihood of a unit escaping from a neighborhood of local minimum. Larger changes in the system's configuration can be obtained due to the unbounded variance of the Cauchy distribution. Noise involved in Cauchy distribution is called colored noise and the noise involved in the Gaussian distribution is called white noise.

2.9.4 Probabilistic Neural Network

Probabilistic neural networks (PNNs) originate in a pattern recognition framework as tools for building classifiers. In this framework the examples of a classification problem are points in a continuous space and they belong to two different classes conventionally named 0 and 1. The PNN was first proposed by Specht (1990). It's a feed-forward neural network, which is widely used in classification and pattern recognition problems. In a PNN algorithm, the parent probability distribution function (PDF) of each class is approximated by a Parzen window and non-parametric function. Then, using the PDF of each class, the class probability of a new input data is estimated and the Bayes rule is applied to allocate the class with the highest posterior probability of new input data. This method minimizes the probability mis-classification. This kind of ANN network is defined by applying a Bayesian network and statistical algorithm called c. It consists of four multilayer, feed-forward network layers: input, hidden, pattern/summation, and output. PNNs are commonly used for classification problems. When an input is present, the first layer computes the distance from the input vector to the training input vectors. This procedure calculates how close the input is to the training input. The second layer sums the contribution from each class of input and produces its net output as a vector of probabilities. Finally, a complete transfer function on the output of the second layer picks the maximum of these probabilities and produces either 1 (positive identification) for that class or 0 (negative) for non-targeted classes.

2.9.5 Cascade Correlation Neural Network

The Cascade Correlation Neural Network (CCNN) has a special network architecture which automatically adapts to the application. It also has a special training process which reduces the computational costs and cures many problems of the back-propagation algorithm at once. First proposed by Fahlman and Lebiere in 1990, the learning algorithm starts with an empty network which has only the input and the output layers and does not have any hidden layers. Because of the absence of hidden neurons, the network can be learned by a simple gradient descent algorithm applied for each output neuron individually. During the learning process, new neurons are added to the network one by one. Each of them is placed into a new hidden layer and connected to all the preceding input and hidden neurons. Once a neuron is finally added to the network (activated), its input connections become frozen and it will not change anymore. The neuron-creation step can be divided into two parts. First, a new, so-called candidate neuron, is connected to all the input and hidden neurons by trainable input connections, but its output is not connected to the network. Then the weights of the candidate neuron can be trained while all the other weights in the network are frozen. The empty layer shows the place of the candidate neuron where it is activated after its input weights are learned and become frozen. Second, the candidate is connected to the output neurons (activated) and then all the output connections (all the input connections of any neuron in the output layer) are trained. The whole process is repeated until the desired network accuracy is obtained.

2.9.6 Cognitive Network

A cognitive network is a data communication network which consists of intelligent devices. By 'intelligent,' we mean that these networks are aware of everything happening inside the device and in the network they are connected to. Using this awareness, they can adjust their operation to match current and near-future network conditions. The cognitive network aims to be proactive so that it can predict most of the usual use cases before they occur and adapt to those beforehand. If predictions fail, it falls back to reactive method and looks for the optimal way of handling the new situation. In any case, the cognitive network learns from every situation it encounters and uses that information for future cases. The main goal of the cognitive network is to increase network efficiency and performance. An important aspect of the cognitive network is that it optimizes data communication for the entire network between the sender and the receiver to meet required end-to-end goals of users of the network. A network becomes cognitive when all the statically configured parts of the network are replaced with self-adjusting and self-aware components. Statically configured nodes are not cognitive, because they need an external operator (human) to make decisions and take care of configuration. The promise of cognitive networking is that the network itself can find optimal ways of connecting devices and tuning network parameters to achieve the best performance for data transfers. It can even optimize for events that have not happened but are likely to happen. The conventional network forwards packets using routing algorithms and detects failures after packets are lost. It also knows the status of every node, so it doesn't send data using a route that cannot deliver the packet, so it prevents congestion. There have been multiple definitions of cognitive networking each one refining and tightening the definition. Thomas et al. define CN as 'a cognitive network is a network with a cognitive process that can perceive current network conditions, and then plan, decide, and act on those conditions.' The network can learn from these adaptations and use them to make future decisions, all while taking into account end-to-end goals.

2.9.7 Cellular Neural Network

A cellular neural network (CNN), also known as cellular nonlinear network, is an array of dynamical systems (cells) or coupled networks with local connections only. Cells can be arranged in several configurations; however, the most popular is the two-dimensional CNNs organized in an eight-neighbor rectangular grid. Each cell has an input, a state, and an output, and it interacts directly only with the cells within its radius of neighborhood r: when r = 1, which is a common assumption, the neighborhood includes the cell itself and its eight nearest neighboring cells. In general, the state of each cell, and, hence its output, depends only on the input and the output of its neighbor cells and the initial state of the network, by varying the values of the connections among cells (i.e., its interaction weights).

2.9.8 Optical Neural Network

Optical neural networks are developed from optical information processing and neural network theory. Biological neural networks function on an electrochemical basis, while optical neural networks use electromagnetic waves. Optical interfaces to biological neural networks can be created ontogenetically, but is not the same as an optical neural network. In biological neural networks there exist a lot of different mechanisms for dynamically changing the state of the neurons; these include short-term and long-term synaptic plasticity. Synaptic plasticity is among the electrophysiological phe-nomena used to control the efficiency of synaptic transmission: long-term for learning and memory and short-term for short transient changes in synaptic transmission efficiency. Implementing this with optical components is dif-ficult and requires advanced photonic materials. Properties that might be desirable in photonic materials for optical neural networks include the abil-ity to change their efficiency of transmitting light, based on the intensity of incoming light.

2.9.9 Summary

In this section, some specific networks based on their special characteris-tics and performance have been discussed. The networks are designed for optimization problems and classifications. Among these, Gaussian machine, Cauchy machine, probabilistic neural networks, cascade correlation net-works, cognition networks, cellular neural networks, and optical neural networks have been mentioned specifically. Cauchy and Gaussian networks are the variation of fixed-weight optimization networks. Probabilistic neu-ral networks are designed using the probability theory to classify the input data. Cascade correlation networks are designed depending on the hierar-chical arrangements of the hidden units. A cellular neural network (CNN), also known as cellular nonlinear network, is an array of dynamical systems (cells) or coupled networks with local connections only.

2.10 Working Principle of ANN

2.10.1 Introduction

Artificial Neural Networks are mathematical inventions inspired by obser-vations made of biological systems. ANN has gained popularity among hydrologists in recent decades due to its large array of applications in the field of engineering research. The first neuron was produced in 1943 by the neurophysiologist Warren McCulloch and the logician Walter Pitts. In 1969 Minsky and Papert wrote a book in which they generalized the limitations of ANNs. The era of renaissance started with John Hopfield in 1984 when he introduced recurrent neural network architecture.

The purpose of ANN is as a mapping function (i.e., mapping an input space to an output space). ANNs have excellent flexibility and high effi-ciency in dealing with nonlinear and noisy data in hydrological modeling. Some of the advantages of using an ANN are input–output mapping, self-adaptivity, real-time operation, fault tolerance, and pattern recognition. A typical ANN consists of a number of nodes that are organized according to a particular arrangement. It consists of 'neurons' which are interconnected computational elements that are arranged in a number of layers which can be single or multiple. Figure 2.6 shows the typical structure of a neuron.

A neural network is a massively parallel distributed processor that has a natural propensity for storing experiential knowledge and making it available for use. It resembles the brain in two respects:

1. Knowledge is acquired by the network through a learning process.
2. Interneuron connection strengths known as synaptic weights are used to store the knowledge.

The human nervous system may be viewed as a three-stage system—receptors, neural networks, and effectors go to work whenever a stimulus is generated followed by a response (the output). Central to the system is the brain, represented by the neural (nerve) net, which continuously receives information, perceives it, and makes appropriate decisions.

Two sets of arrows are shown in the block diagram (Figure 2.2) of the nervous system. Those pointing from left to right indicate the forward transmission of information-bearing signals through the system. The arrows pointing from right to left signify the presence of feedback in the system. The receptors convert stimuli from the human body or the external environment into electrical impulses that convey information to the neural net (the brain). The effectors convert electrical impulses generated by the neural net into discernible responses as system outputs.

A typical ANN consists of a number of nodes that are organized according to a particular arrangement. It consists of 'neurons' which are interconnected computational elements that are arranged in a number of layers which can

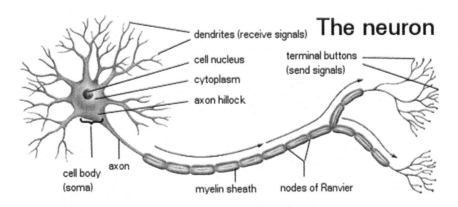

FIGURE 2.6
A typical structure of a neuron.

be single or multiple. A neuron is an information-processing unit that is fundamental to the operation of a neural network. Each pair of neurons is linked and is associated with weights.

The three basic elements of the neuronal model are:

1. **Synapses:** A set of synapses or connecting links, each of which is characterized by its own weight or strength. Specifically, a signal x_j at the input of synapse j connected to neuron k is multiplied by synaptic weight w_{kj}. Here the subscript 'k' refers to the neurons in question and the subscript 'j' refers to the input end of the synapse to which the weight refers. Unlike a synapse in the brain, the synaptic weight of an artificial neuron may lie in a range that includes negative as well as positive values.

   The block diagram in Figure 2.7 shows the model of a neuron which forms the basis for designing (artificial) neural networks.

2. **Adder:** This is used for summing weight of artificial neurons, which may lie in a range that includes negative as well as positive values.

3. **Activation Function:** Used for limiting the amplitude of the output of a neuron. The activation function is also referred to as a squashing function in that it squashes (limits) the permissible amplitude range of the output signal to some finite value. Typically, the normalized amplitude range of the output of a neuron is written as the closed unit interval [0, 1] or alternatively [−1, 1]. The neuronal model as shown in Figure 2.7 also includes an externally applied *bias* denoted by b_k. The bias b_k has the effect of increasing or lowering the net input of the activation function, depending on whether it is positive or negative, respectively.

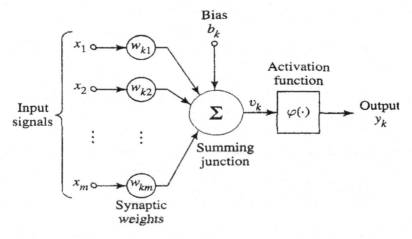

FIGURE 2.7
Nonlinear model of a neuron.

In mathematical terms, a neuron 'k' can be written in the form of the equations:

$$u_k = \sum_{j=1}^{m} w_{kj} x_j \tag{2.1}$$

and

$$y_k = \varphi(u_k + b_k) \tag{2.2}$$

Where x_1, x_2, ..., x_m are the input signals; w_{k1}, w_{k2},..., w_{km} are the synaptic weights of neuron k; u_k is the linear combiner output due to the input signals; b_k is the bias; $\varphi(\bullet)$ is the activation function; and 'y_k' is the output signal of the neuron. The use of bias b_k has the effect of applying an affine transformation of the output u_k of the linear combiner in the model of figure and is shown, by

$$V_k = u_k + b_k \tag{2.3}$$

Depending upon whether the bias b_k is positive or negative, the relationship between the induced local field or activation potential v_k of neuron k and linear combiner output u_k is modified.

2.10.2 Types of Activation Function

The behavior of an ANN depends on both the weights and the input–output function (transfer function) that is specified for the units (Figure 2.8)
   This function typically falls into one of three categories:

1. **Linear (or ramp):** The output activity is proportional to the total weighted output.
2. **Threshold:** The output is set at one of two levels, depending on whether the total input is greater than or less than some threshold value.
3. **Sigmoid:** The output varies continuously but not linearly as the input changes. Sigmoid units bear a greater resemblance to real neurons than do linear or threshold units, but all three must be considered rough approximations.

2.10.3 ANN Architecture

ANN architecture consists of three layers (i.e., the input layer, the hidden layer, and the output layer). The network consists of three distinctive modes: training, cross-validation, and testing. The behavior of an ANN depends on both the weights and the input–output function (transfer function) that is specified for the units. This function typically falls into one of three categories: linear (or ramp), threshold, or sigmoid. An important step in developing

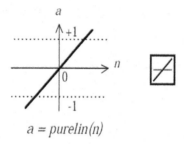

$$a = purelin(n)$$

Linear Transfer Function

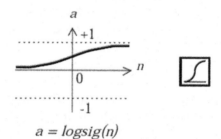

$$a = logsig(n)$$

Log-Sigmoid Transfer Function

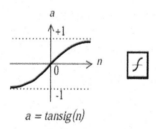

$$a = tansig(n)$$

Tan-Sigmoid Transfer Function

FIGURE 2.8
Types of activation function.

an ANN model is the determination of its weight matrix through training. There are primarily two types of training mechanisms, supervised and unsupervised.

1. **Supervised Training**

   A supervised training algorithm requires an external teacher to guide the training process. The primary goal in supervised training is to minimize the error at the output layer by searching for a set of connection strengths that cause the ANN to produce outputs that are equal to or closer to the targets. A supervised training mechanism called back-propagation training algorithm is normally adopted in most of the engineering applications. ANN is trained by adjusting the values of these connection weights between network elements. The weighted inputs in each layer are processed from neurons in the previous layer and its output transmitted to neurons in the next layer. A transfer function is used to convert a weighted function of the input to get the output.

2. **Unsupervised Training**

Another class of ANN models that employ an 'unsupervised training method' is called self-organizing neural networks. The data passing through the connections from one neuron to another are multiplied by weights that control the strength of a passing signal. When these weights are modified, the data that is transferred through the network changes; consequently, the network output also changes. The signal emanating from the output node(s) is the network's solution to the input problem. Each neuron multiplies every input by its interconnection weight, sums the product, and then passes the sum through a transfer function to produce its result. This transfer function is usually a steadily increasing S-shaped curve, called a sigmoid function.

2.10.4 Learning Process

Learning is a process by which the free parameters of a neural network are adapted through a continuous process of stimulation by the environment in which the network is embedded. The type of learning is determined by the manner in which the parameter changes take place. This definition of the learning process implies the following sequence of events:

1. The neural network is stimulated by an environment.
2. The neural network undergoes changes as a result of this stimulation.
3. The neural network responds in a new way to the environment because of the changes that have occurred in its internal structure.

Let $w_{kj}(n)$ denote the value of the synaptic weight w_{kj} at time n. At time n, an adjustment $\Delta w_{kj}(n)$ is applied to the synaptic weight $w_{kj}(n)$, yielding the updated value.

$$W_{KJ}(N+1) = W_{KJ}(N) + \Delta W_{KJ}(N) \qquad (2.4)$$

A prescribed set of well-defined rules for the solution of a learning problem is called a learning algorithm. As one would expect, there is no unique learning algorithm for the design of neural networks. Rather, we have a 'kit of tools' represented by a diverse variety of learning algorithms, each of which offers advantages of its own. Basically, learning algorithms differ from each other in the various ways in which the adjustment Δw_{kj} to the synaptic weight w_{kj} is formulated.

2.10.5 Feed-Forward Back Propagation

Multilayer perceptrons have been applied successfully to solve some difficult and diverse problems by training them in a supervised manner with a highly popular algorithm known as the error back-propagation algorithm. This algorithm is based on the error-correction learning rule. Basically, the error back-propagation process consists of two passes through the different layers of the network: a forward pass and a backward pass. In the forward pass, activity pattern (input vector) is applied to the sensory nodes of the network, and its effect propagates through the network, layer-by-layer. Finally, a set of outputs is produced as the actual response of the network.

During the forward pass the synaptic weights of the network are all fixed. During the backward pass, on the other hand, the synaptic weights are all adjusted in accordance with the error-correction rule. Specifically, the actual response of the network is subtracted from a desired (target) response to produce an error signal. This error signal is then propagated backward through the network against the direction of synaptic connections—hence the name 'error backpropagation'. The synaptic weights are adjusted so as to make the actual response of the network move closer to the desired response. The error back-propagation algorithm is also referred to in literature as the back-propagation algorithm, or simply back-prop. The feed-forward back-propagation neural network in Figure 2.9 is fully connected, which means that a neuron in any layer is connected to all neurons in the previous layer. Signal flow through the network progresses in a forward direction, from left to right and on a layer-by-layer basis. The connection weights manifest the importance of input to the overall estimation process. The fitting error between the desired and estimated output is used as feedback to enhance the performance of the network by altering the connection weights:

$$\text{Error} = \sum_{j=1}^{N} (y_j - d_j)^2 \qquad (2.5)$$

Where, N = number of output nodes, y_j = calculated output, and d_j = desired data value.

This process is repeated until establishing a successive layer. Therefore, these kinds of networks are called feed-forward backpropagation (FFBP) networks, which are the most popular supervised algorithm for training networks in prediction, pattern recognition, and nonlinear function fitting. Training (calibrating) is a crucial process in which the network is tested by a set of data pairs (input–output) and the initial conditions changed in each

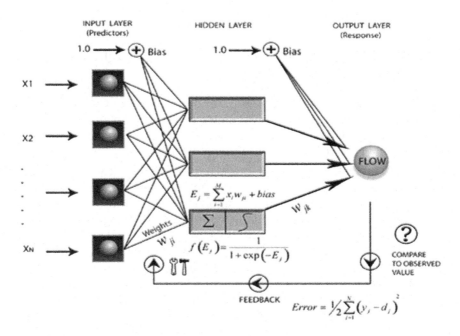

FIGURE 2.9
Three-layered FFNN with BP training algorithm.

iteration step to achieve accurate forecasting. Minimization is performed by calculating the gradient for each node at the output layer.

$$\delta_k = d\sigma_k \cdot (y_k - d_k) \tag{2.6}$$

d_k = the derivative of the sigmoid function applied at y_k which is defined for each kth output node. For the hidden layer (one layer back), the gradient function becomes

$$\delta_j = d\sigma_j \cdot \sum_{i=1}^{N} \delta_j W_{jk} \tag{2.7}$$

Where d_j = the derivative of the sigmoid function and w_{jk} = weight value from hidden node j to output node k. When the input data are chosen, then the network runs; the weights for each connection are updated until the error is minimized to a predefined error target or the desired number of training periods is reached:

$$\Delta W_{jk} = W_{jk} - \eta \delta_k y_j \tag{2.8}$$

Where 'η' is the learning rate of each layer back to the network. Each pass through the training data is called an epoch. In the MATLAB routines, the user can define the number of epochs prior to analysis and manually adjust until the plausible performance is achieved in the trial and error period.

The use of neural networks offers the following useful properties and capabilities:

1. **Nonlinearity:** A neuron is basically a nonlinear device. Consequently, a neural network, made up of an interconnection of neurons, is itself nonlinear. Moreover, the nonlinearity is of a special kind in the sense that it is distributed throughout the network.

2. **Input–output mapping:** A popular paradigm of learning called supervised learning involves the modification of the synaptic weights of a neural network by applying a set of training samples. Each sample consists of a unique input signal and the corresponding desired response. The network is presented, a sample is picked at random from the set, and the synaptic weights (free parameters) of the network are modified so as to minimize the difference between the desired response and the actual response of the network produced by the input signal in accordance with an appropriate criterion. The training of the network is repeated for many samples in the set until the network reaches a steady-state, where there are no further significant changes in the synaptic weights. The previously applied training samples may be re-applied during the training session, usually in a different order. Thus, the network learns from the samples by constructing an input–output mapping for the problem at hand.

3. **Adaptability:** Neural networks have a built-in capability to adapt their synaptic weights to changes in the surrounding environment. In particular, a neural network trained to operate in a specific environment can be easily retrained to deal with minor changes in the operating environmental conditions. Moreover, when it is operating in a non-stationary environment, a neural network can be designed to change its synaptic weights in real time. The natural architecture of a neural network for pattern classification, signal processing, and control applications coupled with the adaptive capability of the network, makes it an ideal tool for use in adaptive pattern classification, adaptive signal processing, and adaptive control.

4. **Contextual information:** Knowledge is represented by the very structure and activation state of a neural network. Every neuron in the network is potentially affected by the global activity of all other neurons in the network. Consequently, contextual information is dealt with naturally by a neural network.

5. **Fault tolerance:** A neural network, implemented in hardware form, has the potential to be inherently 'fault tolerant' in the sense that its performance is degraded gracefully under adverse operating (e.g., if

a neuron or its connecting links are damaged, the recall of a stored pattern is impaired in quality). However, owing to the distributed nature of information in the network, the damage has to be extensive before the overall response of the network is degraded seriously. Thus, in principle, a neural network exhibits a graceful degradation in performance rather than catastrophic failure.

6. **VLSI Implementability:** The massively parallel nature of a neural network makes it potentially fast for the computation of certain tasks. This same feature makes a neural network ideally suited for implementation using very-large-scale-integrated (VLSI) technology.

7. **Uniformity of analysis and design:** Basically, neural networks enjoy universality as information processors. We say this in the sense that the same notation is used in all the domains involving the application of neural networks. This feature manifests itself in different ways:

   a) Neurons, in one form or another, represent an ingredient common to all neural networks.

   b) This commonality makes it possible to share theories and learning algorithms in different applications of neural networks.

   c) Modular networks can be built through seamless integration of modules.

8. **Neurobiological analogy:** The design of a neural network is motivated by analogy with the brain, which is living proof that fault-tolerant parallel processing is not only physically possible but also fast and powerful. Neurobiologists look to (artificial) neural networks as a research tool for the interpretation of neurobiological phenomena. On the other hand, engineers look to neurobiology for new ideas to solve problems more complex than those based on conventional hardwired design techniques. The neurobiological analogy is also useful in another important way: It provides hope and belief that physical understanding of neurobiological structures could influence the art of electronics and thus VLSI.

2.10.6 Strengths of ANN

1. ANNs are better in terms of result accuracy than almost all prevalent analytical, statistical, or stochastic schemes (Jain and Deo, 2004).

2. ANNs methodologies have been reported to have the capability of adapting to a nonlinear and multivariate system having complex interrelationships which may be poorly defined and not clearly understood using mathematical equations (Thirumalah and Deo, 1998).

3. Input data that are incomplete and ambiguous or data with noise, can be handled properly by ANNs because of their parallel processing (ASCE, 2000a; Flood and Kartam-I, 1993).

4. ANNs are able to recognize the relation between the input and output variables without explicit physical consideration of the system or knowing underlying principles because of the generalizing capabilities of the activation function (ASCE, 2000a; Thirumalah and Deo, 1998).

5. Accuracy of ANNs increases as more and more input data is made available to it (Tokar and Markus, 2000).

6. Time is consumed in arriving at best network and training, but ANNs once trained, are easy to use. It is much faster than a physical-based model which it approximates (ASCE, 2000a).

7. ANNs are able to adapt to solutions over time to compensate for changing circumstances (suitable for time-variant problems).

8. ANNs are more suitable for dynamic forecasting problems because the weights can be updated when fresh observations are made available (Thirumalah and Deo, 1998).

9. Neural networks can be complementary or alternative to many complex numerical schemes including FEM/FDM (Jain and Deo, 2004).

2.10.7 Weaknesses of ANN

1. ANN's extrapolation capabilities, beyond its calibration range, are not reliable. During prediction ANN is likely to perform poorly if it faces inputs that are far different from the examples it is exposed to during training. Therefore, prior information of the system is of utmost importance to obtain reasonably accurate estimates (ASCE, 2000a).

2. It is not always possible to determine the significance of the input variables prior to the exercise and it is important to identify and eliminate redundant input variables that do not make a significant contribution to the model. This would result in a more efficient model.

3. The knowledge contained in the trained networks is difficult to interpret because it is distributed across the connection weights in a complex manner.

4. The success of ANN application depends both on the quality and quantity of data available (ASCE, 2000a), type and structure of the neural network adopted, and method of training (Flood and Kartam-II, 1993).

5. Determining the ANN architecture is problem-dependent trial and error process (Shigdi and Gracia, 2003). The choice of network architecture, training algorithm, and definition of error are usually determined by the user's past experience and preference, rather than the physical aspects of the problem (ASCE, 2000a).

6. Initialization of weights and threshold values are an important consideration (Kao, 1996). This problem is faced particularly while implementing the back-propagation training algorithm. Some researchers have tried to overcome this problem by using Genetic Algorithm (GA) global search method.

7. While training the network, there is a danger of reaching local optimum especially for the back-propagation algorithm. Global search techniques like GA and simulated annealing are useful in such conditions.

8. Representing temporal variations is often achieved by including past inputs/outputs as current inputs. However, it is not immediately clear how far back one must go in the past to include temporal effects.

2.10.8 Working of the Network

A typical neural network represents the interconnection of computational elements called neurons or nodes, each of which basically carries out the task of combining the input, determining its strength by comparing the combination with a bias (or alternatively passing through a nonlinear transfer function), and firing out the result in proportion to such a strength.

Mathematically,

$$O = \frac{1}{1 + e^{-s}} \tag{2.9}$$

$$S = \left(x_1 w_1 + x_2 w_2 + x_3 w_3 + \dots\right) + \theta \tag{2.10}$$

Where O is the output from a neuron; x_1, x_2, ..., are the input values; w_1, w_2, ..., are the weights along with the linkages connecting two neurons, indicating strength of the connections; and θ is the bias value. The equation above indicates a transfer function of a sigmoid nature commonly used, although there are other forms available, like sinusoidal, Gaussian, and hyperbolic tangent. A majority of the applications made in ocean engineering so far have involved a feed-forward type of network as opposed to a feedback or recurrent one. A feed-forward multilayer network would consist of an input layer, one or more hidden layers, and an output layer of neurons as shown in Figure 2.9. The mathematical functioning of such a network is given in a four-step procedure:

1. Sum up weighted inputs, that is

$$Nod_j = \sum_{i=1}^{NIN} \left(w_{ij} x_i\right) + \theta_j$$

Where, Nod_j = summation for jth hidden node; NIN = total number of input nodes; W_{ij} = connection weight ith input and jth hidden node; x_i=normalized input at $_i$th input node; θ=bias value at $_j$th hidden node.

2. Transform the weighted input:

$$Out_j = \frac{1}{1+e^{-Nod_j}}$$

Where Out_j= output from $_j$th hidden node.

3. Sum up the hidden node outputs:

$$Nod_k = \sum_{J=1}^{NHN} \left(w_{ik}Out_j\right)+ \theta_k$$

Where Nod_k = summation for kth output node; NHN = total number of hidden nodes; w_{ik} = connection weight between jth hidden and kth output node; k = bias at kth output node.

4. Transform the weighted sum:

$$Out_k = \frac{1}{1+e^{-Nod_j}}$$

Where Out_k = output at $_k$th output node.

Examples are first used to train a network and the strengths of interconnections (or weights) are accordingly fixed. Thereafter, the network becomes ready for application to any unseen real-world example. A supervised type of training involves feeding input–output examples until the network develops its generalization capability while an unsupervised training would involve classification of the input into clusters by some rule. In ocean engineering applications, it is the supervised learning that is most common. During such training, the network output is compared with the desired or actual targeted one and the resulting error or difference is processed through some training algorithm wherein the connection weights and bias are continuously changed by following an interactive process until the desired error tolerance is achieved. The most common training algorithm is the standard backpropagation, although numerous training schemes are available to impart better training with the same set of data.

The overall objective of a training algorithm is to reduce the global error, E, defined below:

$$E = \frac{1}{P}\sum_{p=1}^{P}E_p$$

Where P = the total number of training patterns; E_p error at $_p$th training pattern given by

$$E_p = \frac{1}{2} \sum_{k=0}^{N} (O_k - t_k)^2$$

Where, N = the total number of output nodes; O_k = output at the kth output node; t_k = target output at the kth output node. During training, the weights over difference linkages as well as the bias values are adjusted until the error Ep reaches its minimum. The back-propagation algorithm minimizes the global error according to the steepest or gradient descent approach where the correction to a weight, Δw, is made proportional to the rate of change of the error with respect to that weight; that is

$$\Delta wa\left(\frac{\partial E}{\partial w}\right)$$

One of the earliest employments of NN in civil engineering studies was due to Flood (1989) and it pertained to construction activities.

2.10.9 Summary

In this section, the working strategy of ANN is discussed. Various activation functions, learning processes, weaknesses, and strengths are explained briefly. Mathematical functions are described. Useful capabilities of ANNs, such as handling nonlinearity, adaptivity, fault tolerance, and analogy, are discussed. The core objective of ANNs (input–output mapping) has been highlighted.

Bibliography

Ackley, D., Hinton, G., and Sejnowski, T. 1985. A learning algorithm for Boltzmann machines, *Cognitive Science*, 9, 147–169.

ASCE Task Committee. 2000a. Artificial neural networks in hydrology-I: preliminary concepts, *Journal of Hydrologic Engineering, ASCE*, 5(2), 115–123.

ASCE Task Committee. 2000b. Artificial neural networks in hydrology-II: hydrologic applications. *Journal of Hydrologic Engineering, ASCE*, 5(2), 124–137.

Garrett, J.H., et al. 1993. "Engineering applications of artificial neural networks." *Journal of Intelligent Manufacturing*, 4, 1–21.

Haykin, S. 1999. *Neural Networks: A Comprehensive Foundation*, 2nd Edn, Prentice-Hall, Englewood Cliffs, NJ.

Hebb, D.O. 1949. *The Organization of Behavior: A Neuropsychological Theory*. Wiley, New York.

Hopfield, J.J. 1982. Neural networks and physical systems with emergent collective computational abilities, *Proceedings of the National Academy of Sciences of the United States of America*, 79, 2554–2558.

Hopfield, J.J. 1984. Neurons with graded response have collective computational properties like those of two-state neurons, *Proceedings of the National Academy of Sciences of the United States of America*, 81, 3088–3092.

Kohonen, T. 1982. Analysis of a simple self-organizing process. *Biological Cybernetics*, 44, 135–140.

Kosko, B. 1988. Bidirectiond associative memories, *IEEE Transactions on Systems, Man, and Cybernetics*, SMC-L8, 49–60.

Linsker, R. 1988. Self-organization in a perceptual network, *Computer*, 21, 105–117.

Masters, T. 1993. *Practical Neural Network Recipes in C++*. Academic Press, San Diego, CA.

McCulloch, W.S. and Pitts, W. 1943. A logical calculus of the ideas immanent in nervous activity, *Bulletin of Mathematical Biology*, 52(1–2), 99–115.

Minsky, M. 1954. Neural Nets and the Brain Model Problem. Doctoral dissertation, Department of Mathematics, Princeton University, Princeton, NJ.

Poggio, T. and Girosi, F. 1990. Regularization algorithms for learning that are equivalent to multilayer networks, *Science*, 247, 978–982.

Rosenblatt, F. 1958. The perceptron: A probabilistic model for information storage and organization in the brain, *Psychological Review*, 65, 386–408.

Rumelhart, D.E., Hiton, G.E., and Williams, R.J. 1986. Learning representations by back-propagating errors, *Nature*, 323, 533–536.

Sayed, T., and Razavi, A. 2000. Comparison of neural and conventional approaches to mode choice analysis, *Journal of Computing in Civil Engineering*, ASCE, 14(1), 23–30.

Shi, J.J. 2000. Reducing prediction error by transforming input data for neural networks, *Journal of Computing in Civil Engineering*, ASCE, 14(2), 119–116.

Widrow, B. and Hoff, M.E. 1960. Adaptive Switching Circuits, 1960 IRE WESCON Convention Record, pp. 96–104.

Zurada, J.M. 1992. *Introduction to Artificial Neural Systems*, West Publishing Company, St. Paul, MN, 1992.

3

Fuzzy Logic

In 1965, Lotfi A. Zadeh of the University of California, Berkeley, published 'Fuzzy Sets,' which laid out the mathematics of fuzzy set theory and, by extension, fuzzy logic. Zadeh had observed that conventional computer logic couldn't manipulate data that represented subjective or vague ideas, so he created fuzzy logic to allow computers to determine the distinctions among data with shades of gray similar to the process of human reasoning.

3.1 Introduction to Classical Sets and Fuzzy Sets

It is instructive to introduce fuzzy sets by first reviewing the elements of classical (crisp) set theory. Crisp sets are a special form of fuzzy sets; they are sets without ambiguity in their membership (i.e., they are sets with unambiguous boundaries). It will be shown that fuzzy set theory is a mathematically rigorous and comprehensive set theory useful in characterizing concepts (sets) with natural ambiguity.

3.1.1 Classical Sets

In the case of crisp relation there are only two degrees of relationship between the elements of sets in a crisp relation: 'completely related' and 'not related.' A crisp relation represents the presence or absence of association, interaction, or interconnectedness between the elements of two or more sets.

Define a universe of discourse, X, as a collection of objects—all having the same characteristics.

The individual elements in the universe X will be denoted as x. The features of the elements in X can be discrete, countable integers, or continuous-valued quantities on the real line.

3.1.2 Fuzzy Sets

A fuzzy set is a set without a crisp, clearly defined boundary. It can contain elements with a partial degree of membership. In other words, a fuzzy set is a set containing elements with varying degrees of membership in the set. A fuzzy set is different from the classical crisp set because members of a crisp set will not be members unless their membership is full in that set. Figure 3.1

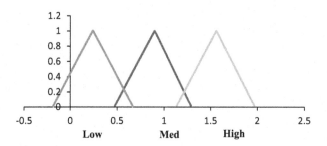

FIGURE 3.1
Fuzzification.

shows fuzzification and Figure 3.2 shows the structure of a fuzzy set. For further details see Dubois and Prade, 1980.

3.1.3 Summary

Fuzzy logic is another form of artificial intelligence, but its history and applications are more recent than artificial intelligence-based expert systems. Fuzzy logic is based on the fact that human thinking does not always follow crispy 'YES–NO' logic; rather, it is often vague, uncertain, and indecisive. Fuzzy logic deals with problems that have vagueness, uncertainty, imprecision, approximations or partial truth or qualitative mess.

3.2 Classical Relations and Fuzzy Relations

Understanding relations is central to the understanding of a great many areas addressed in this textbook. Relations are intimately involved in logic,

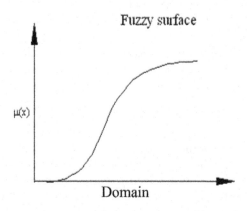

FIGURE 3.2
Structure of fuzzy set.

approximate reasoning, rule-based systems, nonlinear simulation, synthetic evaluation, classification, pattern recognition, and control. Relations will be referred to repeatedly in this text in many different applications areas. Relations represent mappings for sets just as mathematical functions do; relations are also very useful in representing connectives in logic.

3.2.1 Introduction

Fuzzy logic is based on fuzzy set theory in which a particular object or variable has a degree of membership in a given set which may be anywhere in the range of 0 to 1. This is different from a conventional set theory based on Boolean logic in which a particular object or variable is either a member (logic 1) of a given set or it is not (logic 0). The basic set operations like union (OR), intersection (AND) and complement (NOT) of Boolean logic are also valid for fuzzy logic.

3.2.2 Classical Relation

Classical (crisp) relations are structures that represent the presence or absence of correlation, interaction, or propinquity between the elements of two or more crisp sets; in this case, a set could also be the universe. There are only two degrees of relationship between elements of the sets in a crisp relation: 'completely related' and 'not related,' in a binary sense.

A subset of the Cartesian product A1 × A2 × … × Ar is called an *r-ary relation* over A1, A2, …, Ar. Again, the most common case is for $r = 2$; in this situation the relation is a subset of the Cartesian product A1 × A2 (i.e., a set of pairs, the first coordinate of which is from A1 and the second from A2). This subset of the full Cartesian product is called a *binary relation from* A1 *into* A2. If three, four, or five sets are involved in a subset of the full Cartesian product, the relations are called ternary, quaternary, and quinary, respectively. In this text, whenever the term *relation* is used without qualification, it is taken to mean a *binary relation*.

The Cartesian product of two universes X and Y is determined as

$$X \times Y = \{(x,y)| \, x \in X, y \in Y\}$$

which forms an ordered pair of every $x \in X$ with every $y \in Y$, forming *unconstrained* matches between X and Y. That is, every element in universe X is related completely to every element in universe Y. The *strength* of this relationship between ordered pairs of elements in each universe is measured by the characteristic function, denoted χ, where a value of unity is associated with *complete relationship* and a value of zero is associated with *no relationship*, i.e.,

$$\chi X \times Y(x,y) = \begin{cases} 1, (x,y) \in X \times Y \\ 0, (x,y) \notin X \times Y \end{cases}$$

3.2.3 Fuzzy Relation

Fuzzy relations also map elements of one universe, say X, to those of another universe, say Y, through the Cartesian product of the two universes. However, the 'strength' of the relation between ordered pairs of the two universes is not measured with the characteristic function, but rather with a membership function expressing various 'degrees' of the strength of the relation on the unit interval [0,1]. Hence, a fuzzy relation R is a mapping from the Cartesian space $X \times Y$ to the interval [0,1], where the strength of the mapping is expressed by the membership function of the relation for ordered pairs from the two universes, or μR (x, y).

Fuzzy composition combines fuzzy relations in different Cartesian product spaces with each other. Similarly, a fuzzy set can also be combined with a fuzzy relation. Out of many different versions of composition operators, the most commonly used operators are max–min operator, max–prod operator and max–average operator.

3.2.4 Tolerance and Equivalence Relations

Let R be the fuzzy relation defined on the set of cities and representing the concept 'very near.' We may assume that a city is certainly (i.e., to a degree of 1) very near to itself. The relation is therefore reflexive. Furthermore, if city A is very near to city B, then B is certainly very near to A. Therefore, the relation is also symmetric. Finally, if city A is very near to city B to some degree, say 0.7, and city B is very near to city C to some degree, say 0.8, it is possible (although not necessary) that city A is very near to city C to a smaller degree, say 0.5. Therefore, the relation is non-transitive.

A fuzzy relation is a fuzzy equivalence relation if all three of the following properties for matrix relations define it:

Reflexivity $\mu R(x_i, y_i) = 1$

Symmetry $\mu R(x_i, y_j) = \mu R(x_j, y_i)$

Transitivity $\mu R(x_i, y_j) = \lambda 1$ and $\mu R(x_j, y_k) = \lambda 2$

$\rightarrow \mu R(x_i, y_k) = \lambda$ where $\lambda \geq \min[\lambda 1, \lambda 2]$.

Tolerance relation has only the properties of reflexivity and symmetry. A tolerance relation, R, can be reformed into an equivalence relation by at most (n − 1) compositions with itself, where n is the number of rows or columns of R.

3.2.5 Summary

In this section, the properties and operations of crisp and fuzzy relations have been discussed. The related concept is used for nonlinear simulation, classification, and control. The composition of relations provides a view of extending fuzziness into functions. Tolerance and equivalence relations are useful for solving similar classification problems.

3.3 Membership Functions

A membership function is a function that defines how each point or object in the universe of discourse is assigned a degree of membership or membership value between 0 and 1. The membership function can be an arbitrary curve that is suitable in terms of simplicity, convenience, speed, and efficiency.

3.3.1 Introduction

Though a membership function can be an arbitrary curve, there are eleven standard membership functions that are commonly used in engineering applications. These membership functions can be built from several basic functions: piecewise linear functions, sigmoid curve,
Gaussian distribution function, and quadratic and cubic polynomial curves. The simplest membership functions can be formed using straight lines: a triangular membership function, which is a collection of three points forming a triangle; or trapezoidal membership function, which has a flat top and is just a truncated triangle curve. These membership functions built out of straight lines have the advantage of simplicity. Various membership functions are presented in Figure 3.3.
There is a wide choice of membership function types when a membership function is to be selected. There is no hard and strict rule regarding the selection of membership functions. The membership functions can be selected to suit the applications in terms of simplicity, convenience, speed, and efficiency.

3.3.2 Features of Membership Function

Since all information contained in a fuzzy set is described by its membership function, it is useful to develop a lexicon of terms to describe various special features of this function. For purposes of simplicity, the functions shown in the following figures will all be continuous, but the terms apply equally for both discrete and continuous fuzzy sets.
The *core* of a membership function for some fuzzy set A is defined as that region of the universe that is characterized by complete and full membership in the set A. That is, the core comprises those elements x of the universe such that $\mu A(x) = 1$.
The *support* of a membership function for some fuzzy set A is defined as that region of the universe that is characterized by nonzero membership in

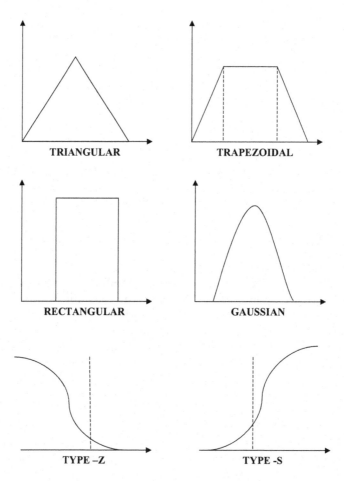

FIGURE 3.3
Various membership functions.

the set A. That is, the support comprises those elements x of the universe such that $\mu A(x) > 0$.

The *boundaries* of a membership function for some fuzzy set A are defined as that region of the universe containing elements that have a nonzero membership but not complete membership. That is, the boundaries comprise those elements x of the universe such that $0 < \mu A\ (x) < 1$. These elements of the universe are those with some *degree* of fuzziness, or only partial membership in the fuzzy set A.

A *normal* fuzzy set is one whose membership function has at least one element x in the universe whose membership value is unity. For fuzzy sets where one and only one element has a membership equal to one, this element is typically referred to as the *prototype* of the set, or the prototypical element.

A *convex* fuzzy set is described by a membership function whose membership values are strictly monotonically increasing, or whose membership values are strictly monotonically decreasing, or whose membership values are strictly monotonically increasing then strictly monotonically decreasing with increasing values for elements in the universe.

3.3.3 Fuzzification

Fuzzification is the process of making a crisp quantity fuzzy. We do this by simply recognizing that many of the quantities that we consider to be crisp and deterministic are actually not deterministic at all; they carry considerable uncertainty. If the form of uncertainty happens to arise because of imprecision, ambiguity, or vagueness, then the variable is probably fuzzy and can be represented by a membership function.

3.3.4 Membership Value Assignment

There are various methods to assign membership values to fuzzy variables. The process of membership value assignment may be by intuition, logical reasoning, procedural method, or algorithmic approach.

- Intuition

  It is based on the common intelligence of humans.

  There should be an in-depth knowledge of the application to which membership value assignment has to be made. The main characteristics of these curves for their usage in fuzzy operations are based on their overlapping capacity.

- Inference

  The inference method uses knowledge to perform deductive reasoning. Forward inference supports the conclusion by deductive reasoning. The knowledge of geometrical shapes and geometry is used for defining membership values. The membership functions may be defined by various shapes: triangular, trapezoidal, bell-shaped, Gaussian, and so on. Other assigning methods are as follows:

- Rank ordering
- Angular fuzzyset
- Neural networks
- Genetic algorithms
- Induction reasoning

3.3.5 Summary

In this section, membership functions and their features have been discussed briefly. The different methods in obtaining membership functions are also explained. The inference method is based on geometrical shapes and geometry, whereas the angular fuzzy set is based on the angular features. Using neural networks and reasoning methods the memberships are tuned in a cyclic fashion and are based on rule structure. The genetic algorithm can provide improvement by using the optimum solution.

3.4 Defuzzification

Mathematically, the defuzzification of a fuzzy set is the process of 'rounding it off' from its location in the unit hypercube to the nearest (in a geometric sense) vertex. If one thinks of a fuzzy set as a collection of membership values, or a vector of values on the unit interval, defuzzification reduces this vector to a single scalar quantity—presumably to the most typical prototype or representative value. Various popular forms of converting fuzzy sets to crisp sets or to single scalar values are introduced later in this section.

3.4.1 Introduction

Fuzzy rule-based systems evaluate linguistic if–then rules using fuzzification, inference, and composition procedures. They produce fuzzy results which usually have to be converted into crisp output. To transform the fuzzy results into crisp, defuzzification is performed. Defuzzification is the process of converting a fuzzified output into a single crisp value with respect to a fuzzy set. The defuzzified value in the fuzzy logic controller (FLC) represents the action to be taken in controlling the process. Some of the commonly used defuzzification methods are discussed here. Centroid method, center of largest area method, height method, first of maxima method, last of maxima method, and mean of maxima method are based on aggregated fuzzy output, that is, all fuzzy outputs corresponding to different rules are aggregated using a union operator (max operator) to an aggregated fuzzy output before defuzzification. The weighted average method and center of sums methods are based on individual output fuzzy sets (Figure 3.4).

3.4.2 Lamda Cut for Fuzzy Sets

Consider a fuzzy set A. The subset will be called 'weak' Lamda cut (or Alfa cut) and is a crisp set of the fuzzy set if it consists of all the elements of a fuzzy set whose membership functions have values greater than or equal to a specified value. On the other hand, a subset is called 'strong' Lamda cut if it consists of all the elements of a fuzzy set whose membership functions have values strictly greater than a specified value. All the Alfa cut sets form a family of crisp sets

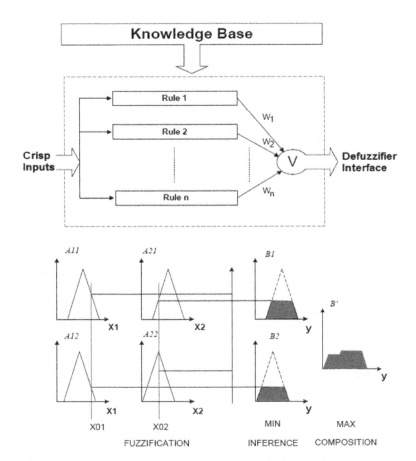

FIGURE 3.4
Aggregation for inputs.

3.4.3 Defuzzification Methods

Defuzzification is the process of transforming the fuzzy quantity into a precise or crisp quantity. The output of a fuzzy process may be the union of two or more fuzzy membership functions defined in the universe of discourse of the output variable.

The fuzzy output process may be linked with many output components, and the membership function representing each component of the output can have any shape.

The defuzzification methods usually used are (Figure 3.5):

- Max–min membership principle
- Centroid method
- Weighted average method

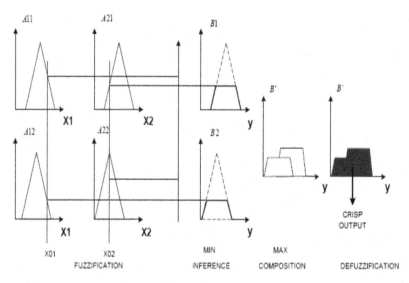

FIGURE 3.5
Fuzzy logic system for defuzzification.

- Mean–max membership
- Center of sums
- Center of the largest area
- First of maxima, last of maxima

Centroid method or center of gravity method is the most commonly used defuzzification method as this method is very accurate and gives smooth output. In the centroid defuzzification method, the defuzzified value is calculated as

$$Z^* = \frac{\int \mu_C(z) \cdot z\,dz}{\int \mu_C(z)\,dz}$$

where Z^* is the defuzzified value, z is the output variable, and $\int \mu_c(z)$ is the membership function of the aggregated fuzzy output. This method is computationally complex.

The center of largest area defuzzification can be used if the aggregated fuzzy set has at least two convex sub-regions (convex region is the region in which the membership values are strictly monotonically increasing or strictly monotonically decreasing or strictly monotonically increasing and then strictly monotonically decreasing with increasing values of points in the universe). Here, the centroid of the convex sub-region with the largest area is determined to obtain the defuzzified value.

$$Z^* = \frac{\int \mu_{Cm}(z) \cdot z \, dz}{\int \mu_{Cm}(z) \, dz}$$

where C_m represents the convex sub-region that has the largest area.

Max-membership method (or height method) is restricted to peaked aggregate membership function. This scheme is given by the mathematical expression

$$\mu_C\left(z^*\right) \geq \mu_C(z) \text{ for all } z \in Z$$

where z^* is the defuzzified output.

First of the maxima method determines the smallest value of the domain of the aggregate membership function having a maximum membership degree in it. Similarly, last of maxima method determines the largest value of the domain of the aggregate membership function having maximum membership degree in it. Mean of maxima method determines the mean of smallest and largest values of the domain of the aggregate membership function having maximum membership degree in it.

Weighted average defuzzification scheme is the most commonly used one in fuzzy logic applications because of its computational efficiency. However, it has the disadvantage that it is limited to symmetrical membership functions. It is given by the mathematical expression

$$Z^* = \frac{\sum \mu_C(\bar{z}) \cdot \bar{Z}}{\sum \mu_C(\bar{z})}$$

where $\bar{z}$ is the centroid of each membership function corresponding to different rules and $\mu_c(\bar{z})$ is the membership value of $\bar{z}$ in that membership function.

Center of sums method is a computationally efficient method applicable to symmetrical or unsymmetrical membership functions. This scheme is given by

$$Z^* = \frac{\int z \cdot \sum_{k=1}^{n} \mu_{C_k}(z) \, dz}{\int \sum_{k=1}^{n} \mu_{C_k}(z) \, dz}$$

Center of sums method is similar to weighted average method as both methods are weighted methods. In the center of sums method, weights are the areas of the membership functions, whereas the weights are membership values in weighted average method.

3.4.4 Summary

In this section, the various methods of defuzzification have been discussed. The defuzzification process is essential in the Mamdani method. The method of defuzzification may be assessed on the basis of the output in the context of the availability of data. Also, Lamda cut fuzzy set concept has been incorporated.

3.5 Fuzzy Arithmetic and Fuzzy Measures

The scientific literature on fuzzy numbers and arithmetic calculations is rich in several approaches to defining fuzzy operations, having many desired properties that are not always present in the implementations of classical extension principle or its approximations (shape preservation, reduction of the overestimation effect, requisite constraints, distributivity of multiplication, division, etc.).

The arithmetic operations and more generally fuzzy calculations are natural when dealing with fuzzy reasoning and systems, where variables and information are described by fuzzy numbers and sets; in particular, procedures and algorithms have to take into account the existing dependencies (and constraints) relating to all the operands involved and their meaning. The essential uncertainties are generally modeled in the preliminary definitions of the variables, but it is very important to pay great attention to how they propagate during the calculations. A solid result in fuzzy theory and practice is that calculations cannot be performed by using the same rules as in arithmetic with real numbers, and, in fact, fuzzy calculus will not always satisfy the same properties (e.g., distributivity, invertibility, and others). If not performed by taking into account existing dependencies between the data, fuzzy calculations will produce excessive propagation of initial uncertainties.

3.5.1 Introduction

In many fields of science (physics, engineering, economics, social, political sciences, etc.) and disciplines, where fuzzy sets and fuzzy logic are applied (e.g., approximate reasoning, image processing, fuzzy systems modeling and control, fuzzy decision-making, statistics, operations research and optimization, computational engineering, artificial intelligence, and fuzzy finance and business) fuzzy numbers and arithmetic play a central role and are frequently, and increasingly, the main instruments. Many studies are reported in this area such as, Kaufmann and Gupta, 1985 and Kim and Mendel, 1995.

The fundamentals of fuzzy arithmetic are nothing but interval arithmetic (in a given closed interval R, how to add, subtract, multiply, and divide).

This interval in R is also called an interval of confidence as it limits the uncertainty of data to an interval.

3.5.2 Fuzzy Arithmetic

If we let A = [a1, a2] and B = [b1, b2] be two closed intervals in R, then we have the following definitions:

Addition and Subtraction

If $x \in [a_1, a_2]$, $y \in [b_1, b_2]$

Then $x + y \in [a_1 + b_1, a_2 + b_2]$ and $x - y \in [a_1 - b_1, a_2 - b_2]$

Therefore $A + B = [a_1, a_2] + [b_1, b_2] = [a_1 + b_1, a_2 + b_2]$.
$$A - B = [a_1, a_2] - [b_1, b_2] = [a_1 - b_2, a_2 - b_1].$$

Image of an Interval

If $x \in [a_1, a_2]$ then its image $-x \in [-a_2, -a_1]$. Therefore, the image of A is denoted by $\bar{A}$ and is defined as

$$\bar{A} = [a_1, a_2] = [-a_2, -a_1]$$

Multiplication

The multiplication of two closed intervals $A = [a_1, a_2]$ and $B = [b_1, b_2]$ of R denoted by A·B is defined as

$$A \cdot B = \left[\min(a_1 b_1, a_1 b_2, a_2 b_1, a_2 b_2), \max(a_1 b_1, a_1 b_2, a_2 b_1, a_2 b_2) \right].$$

Scalar Multiplication and Inverse

Let $A = [a_1, a_2]$ be a closed interval in R and $k \in R$ identifying the scalar k as the closed interval [k, k], the scalar multiplication $k \cdot A$ is defined as

$$k \cdot A = [k, k] \cdot [a_1, a_2] = [ka_1, ka_2]$$

Then $A = [a_1, a_2] \in R$ if $x \in [a_1, a_2]$ and if $0 \notin [a_1, a_2]$ then $\dfrac{1}{x} \in \left[\dfrac{1}{a_2}, \dfrac{1}{a_1} \right]$.

Therefore the inverse of A is denoted by A^{-1} and it is defined as

$$A^{-1} = [a_1, a_2]^{-1} = \left[\frac{1}{a_2}, \frac{1}{a_1} \right] \text{ provided } 0 \notin [a_1, a_2].$$

Division

The division of two closed intervals $A = [a_1, a_2]$ and $B = [b_1, b_2]$ of R denoted by A/B is defined as multiplication $[a_1, a_2]$ and $\left[\dfrac{1}{b_2}, \dfrac{1}{b_1}\right]$ provided $0 \notin [b_1, b_2]$.

Therefore $A/B = [a_1, a_2] / [b_1, b_2]$

$$= [a_1, a_2]\left[\frac{1}{b_2}, \frac{1}{b_1}\right]$$

$$= \left[\min\left(\frac{a_1}{b_2}, \frac{a_1}{b_1}, \frac{a_2}{b_2}, \frac{a_2}{b_1}\right), \max\left(\frac{a_1}{b_2}, \frac{a_1}{b_1}, \frac{a_2}{b_2}, \frac{a_2}{b_1}\right)\right].$$

Max ∨ and min ∧ operations

Let $A = [a_1, a_2]$ and $B = [b_1, b_2]$ be two closed intervals in R then the max ∨ and min ∧ operations on A and B are defined as

$$A \vee B = [a_1, a_2] \vee [b_1, b_2] = [a_1 \vee b_1, a_2 \vee b_2].$$

$$A \wedge B = [a_1, a_2] \wedge [b_1, b_2] = [a_1 \wedge b_1, a_2 \wedge b_2].$$

3.5.3 Fuzzy Extension

Standard arithmetic and algebraic operations, which are based after all on the foundations of classical set theory and can be extended to fuzzy arithmetic and fuzzy algebraic operations. This extension is accomplished with Zadeh's extension principle (Zadeh, 1975). Fuzzy numbers are used here because such numbers are the basis for fuzzy arithmetic. In this context the arithmetic operations are not fuzzy; the numbers on which the operations are performed are fuzzy and, hence, so too are the results of these operations. Conventional interval analysis is reviewed as a prelude to some improvements and approximations to the extension principle, most notably the fuzzy vertex method and its alternative forms.

In engineering, mathematics, and the sciences, functions are ubiquitous elements in modeling. Consider a simple relationship between one independent variable and one dependent variable. This relationship is a single-input, single-output process where the transfer function represents the mapping provided by the general function *f*. In the typical case, *f* is of the analytic form (e.g., $y = f(x)$, the input, x, is deterministic, and the resulting output, y, is also deterministic).

How can we extend this mapping to the case where the input, x, is a fuzzy variable or a fuzzy set, and where the function itself could be fuzzy? That is, how can we determine the fuzziness in the output, y, based on either a fuzzy input or a fuzzy function or both (mapping)? An extension principle developed by Zadeh (1975) and later elaborated by Yager (1986) enables us to extend the domain of a function on fuzzy sets.

The material of the next several sections introduces the extension principle by first reviewing theoretical issues of classical (crisp) transforms, mappings, and relations. The theoretical material then moves to the case where the input is fuzzy but the function itself is crisp, then to the case where the input and the function both are fuzzy.

3.5.4 Fuzzy Measures

The concept of measure is one of the most important concepts in mathematics, as well as the concept of integral respect to a given measure. The classical measures are supposed to hold the additive property. Additivity can be very effective and convenient in some applications, but can also be somewhat inadequate in many reasoning environments of the real world (as in approximate reasoning, fuzzy logic, artificial intelligence, game theory, decision-making, psychology, economy, data mining, etc.) which require the definition of non-additive measures and a large amount of open problems. For example, the efficiency of a set of workers is being measured, the efficiency of the same people doing teamwork is not the addition of the efficiency of each individual working on their own. The concept of fuzzy measure does not require additivity, but it requires monotonicity related to the inclusion of sets. The concept of fuzzy measure can also be generalized by new concepts of measure that pretend to measure a characteristic that is not really related to the inclusion of sets. However, those new measures can show that 'x has a higher degree of a particular quality than y' when x and y are ordered by a preorder (not necessarily the set inclusion preorder). For works on this subject please see Kosko, 1992, 1994 and Schweizer and Sklar, 1963.

3.5.5 Measure of Fuzziness

Fuzziness is a concept that can be only attributed to, and hence arises with, fuzzy sets. As pointed out by G.J. Klir in Chapter 8, 'The question of how to measure fuzziness is one of the fundamental issues of fuzzy set theory.' However, for that purpose, it is first necessary to know what we mean by fuzziness.

3.5.6 Summary

In this section, fuzzy arithmetic, fuzzy extension, fuzzy measures, and the measure of fuzziness have been discussed. The methodology for extending crisp concepts to address fuzzy quantities, such as real algebraic operations on fuzzy numbers, has been discussed. Also, the fuzzy extension principle that linked nonfuzzy elements to fuzzy entities is incorporated. The fuzzy measures concept and the satisfied criteria by a set function was also discussed. The measure of fuzziness linked with uncertainty related to vagueness was briefly explained.

3.6 Fuzzy Rule Base and Approximate Reasoning

3.6.1 Introduction

Fuzzy logic addresses problems in the following ways: first, the meaning of an imprecise proposition is represented as an elastic constraint on a variable; and second, the answer to a query is deduced through a propagation of elastic constraints.

By approximate reasoning, we mean a type of reasoning which is neither very exact nor very inexact. In other words, it is the process or processes by which a possible imprecise conclusion is deduced from a collection of imprecise premises. Simply put, we can say that fuzzy logic plays a key role in approximate reasoning.

3.6.2 Fuzzy Proposition

To extend the reasoning capability, fuzzy logic utilized fuzzy predicates, fuzzy predicate modifiers, fuzzy quantities, and fuzzy qualifiers in the fuzzy propositions. It is only fuzzy propositions that make the fuzzy logic separate from classical logic. Fuzzy propositions are as follows:

- **Fuzzy predicates:** The predicates can be fuzzy such as low, medium, or high.
- **Fuzzy predicate modifiers:** Modifiers may be acting as hedges such as very low, slightly high. These modifiers are necessary to closely specify a variable.
- **Fuzzy quantifiers:** Fuzzy quantifiers are used, like several, most, and many in fuzzy logic (e.g., many people are educated). Fuzzy quantifiers can be used to represent the meaning of propositions linked with probabilities.
- **Fuzzy qualifiers:** It may be truth qualification: Runoff is high (not very true).
- **Probability qualification:** Runoff is high (likely to be).
- **Possibility qualification:** Runoff is high (almost impossible).
- **Usuality qualification:** Runoff is high (usually true during a monsoon).

3.6.3 Formation of Rules

Rules are the essence of fuzzy logic systems. It is working under IF–THEN conditions which are used widely.

IF antecedent, THEN consequent;

IF temperature is high, THEN climate is hot.

There may be unconditional assignment statements also available. Generally, both conditional and unconditional statements are linked with some restrictions on the consequent part of the rule-based system. Fuzzy sets and relations usually model the restrictions. The restrictive statements are usually connected by linguistic connectives like AND, OR, and ELSE.

3.6.4 Decomposition of Rules

Decomposition of rules starts when a compound rule structure is to be reduced to a number of simple rule forms. Compound rules are the result of the combination of simple rules. These rules are usually based on natural language representations.

IF A, THEN B, ELSE C

can be decomposed as

IF A, THEN B, OR IF NOT A, THEN C.

3.6.5 Aggregation of Fuzzy Rules

Aggregation of rules is the process of determining the overall consequences of the individual consequents given by each rule.

3.6.6 Fuzzy Reasoning

Fuzzy reasoning, sometimes called approximate reasoning, includes categorical, qualitative, syllogistic, and dispositional reasoning. In 1979, Zadeh introduced the theory of approximate reasoning. This theory provides a powerful framework for reasoning in the face of imprecise and uncertain information. Central to this theory is the representation of propositions as statements assigning fuzzy sets as values to variables. Suppose we have two interactive variables, $x \in X$ and $y \in Y$, and the causal relationship between x and y is completely known. Namely, we know that y is a function of x, that is $y = f(x)$. Then we can make inferences easily '$y = f(x)$' & '$x = x1$' $\longrightarrow$ '$y = f(x1)$.'

This inference rule says that if we have $y = f(x)$, for all $x \in X$ and we observe that $x = x1$, then y takes the value $f(x1)$. More often than not, we do not know the complete causal link, f, between x and y, we only know the values of $f(x)$ for some particular values of x, that is

>1: If $x = x1$, then $y = y1$
>2: If $x = x2$, then $y = y2$ ...
>n: If $x = xn$, then $y = yn$

If we are given an $x' \in X$ and want to find a $y' \in Y$ which corresponds to x' under the rule base $>$ = {>1,..., >m}, then we have an interpolation problem.

3.6.7 Fuzzy Inference System

Fuzzy inference (reasoning) is the actual process of mapping from a given input to an output using fuzzy logic. The process involves all the pieces that we have discussed in the previous sections: membership functions, fuzzy logic operators, and if–then rules.

Fuzzy inference systems have been successfully applied in fields such as automatic control, data classification, decision analysis, expert systems, and computer vision. Because of its multidisciplinary nature, the fuzzy inference system is known by a number of names, such as fuzzy rule-based system, fuzzy expert system, fuzzy model, fuzzy associative memory, fuzzy logic controller, and simply fuzzy system.

The steps of fuzzy reasoning (inference operations upon fuzzy IF–THEN rules) performed by FISs are:

- Compare the input variables with the membership functions on the antecedent part to obtain the membership values of each linguistic label. (This step is often called fuzzification.)
- Combine (usually multiplication or min) the membership values on the premise part to get firing strength (degree of fulfillment) of each rule.
- Generate the qualified consequents (either fuzzy or crisp) or each rule depending on the firing strength.
- Aggregate the qualified consequents to produce a crisp output. (This step is called defuzzification.)

The rule base and the database are jointly referred to as the knowledge base and include

- a rule base containing a number of fuzzy IF–THEN rules;
- a database which defines the membership functions of the fuzzy sets used in the fuzzy rules.

3.6.7.1 Fuzzy Inference Methods

The most important two types of fuzzy inference methods are Mamdani and Sugeno fuzzy inference methods; Mamdani fuzzy inference is the most commonly used inference method. This method was introduced by Mamdani and Assilian (1975). Another well-known inference method is the so-called Sugeno (or Takagi–Sugeno–Kang) method of fuzzy inference process. This method was introduced by Sugeno (1985) and is also known as the TSK method. The main difference between the two methods lies in the consequent of fuzzy rules.

3.6.8 Fuzzy Expert System

Once the problem has been clearly specified, the process of constructing the fuzzy expert system can begin. Invariably, some degree of data preparation and preprocessing is required. The first major choice the designer has to face is whether to use the Mamdani inference method or the TSK method. The essential difference in these two methodologies is that the result of Mamdani inference is one or more fuzzy sets which must (almost always) then be defuzzified into one or more real numbers, whereas the result of TSK inference is one or more real functions which may be evaluated directly. Thus, the choice of inference methodology is linked to the choice of defuzzification method. Once the inference methodology and defuzzification method have been chosen, the process of enumerating the necessary linguistic variables can commence. This should be relatively straightforward if the problem has been well specified and is reasonably well understood. If this is not the case, then the decision to construct a fuzzy expert system may not be appropriate. The next stage of deciding the necessary terms with their defining membership functions and determining the rules to be used is far from trivial however. Indeed, this stage is usually the most difficult and time-consuming of the whole process (Giarratono and Riley, 1993).

After a set of fuzzy membership functions and rules has been established, the system may be evaluated, usually by comparison of the obtained output against some desired or known output using some form of error or distance function. However, it is very rare that the first system constructed will perform at an acceptable level. Usually, some form of optimization or performance tuning of the system will need to be undertaken. Again, there are a multitude of options that a designer may consider for model optimization. A primary distinction illustrated in Figure 3.6 is the use of either parameter

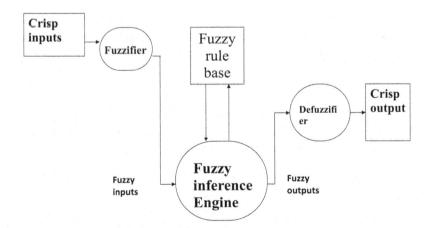

FIGURE 3.6
Structure of a fuzzy inference system.

optimization (in which usually only aspects of the model such as the shape and location of membership functions and the number and form of rules are altered) or structure optimization (in which all aspects of the system, including items such as the inference methodology, defuzzification method, or number of linguistic variables may be altered). In general, though, there is no clear distinction. Some authors consider rule modification to be structure optimization, while others parameterize the rules.

3.6.9 Summary

In this section, various forms of fuzzy propositions are presented. The formation of rules and the decomposition of rules were discussed. The aggregation of fuzzy rules, fuzzy reasoning, fuzzy inference, and fuzzy expert systems were also discussed. The Mamdani and Sugeno FIS provides a base for building a fuzzy rule-based system. The fuzzy expert system overview was provided, which deals with certainty.

3.7 Fuzzy Decision-Making

3.7.1 Introduction

Decision-making is an important scientific, social, and economic endeavor. To be able to make consistent and correct choices is the essence of any decision process imbued with uncertainty. Most issues in life, as trivial as we might consider them, involve decision processes of one form or another.

From the moment we wake in the morning to the time we place our bodies at rest at the day's conclusion, we make many, many decisions. What should we wear for the day; should we take an umbrella; what should we eat for breakfast, for lunch, for dinner; should we stop by the gas station on the way to work; what route should we take to work; should we attend that seminar at work; should we write the memorandum to our colleagues before we make the reservations for our next trip out of town; should we go to the store on our way home; should we take the kids to that new museum before, or after, dinner; should we watch the evening news before retiring; and so on and so forth.

The problem in making decisions under uncertainty is that the bulk of the information we have about the possible outcomes, about the value of new information, about the way the conditions change with time (dynamic), about the utility of each outcome–action pair, and about our preferences for each action is typically vague, ambiguous, and otherwise fuzzy. In some situations, the information may be robust enough so that we can characterize it with probability theory.

Steps for Decision-Making

Let us now discuss the steps involved in the decision-making process:

1. **Determine the set of alternatives:** In this step, the alternatives from which the decision has to be taken must be determined.
2. **Evaluate the alternatives:** Here, the alternatives must be evaluated so that the decision can be taken about one of the alternatives.
3. **Compare the alternatives:** In this step, a comparison between the evaluated alternatives is done.

3.7.2 Individual and Multi-Person Decision-Making

In individual decision-making, only a single person is responsible for making decisions. This decision-making model can be characterized as:

- Set of possible actions
- Set of goals $G_i(i \in X_n)$; $G_i(i \in X_n)$
- Set of Constraints $C_j(j \in X_m)$; $C_j(j \in X_m)$

The goals and constraints stated above are expressed in terms of fuzzy sets. Now consider a set A. Then, the goals and constraints for this set are given by

$$G_i(a) = \text{composition} \left[G_i(a) \right] = G_i^1 \left(G_i(a) \right) \text{ with } G_i^1$$

$$C_j(a) = \text{composition} \left[C_j(a) \right] = C_j^1 \left(C_j(a) \right) \text{ with } C_j^1 \text{ for } a \in A$$

The fuzzy decision in the above case is given by

$$F_D = \min \left[i \in X_n^{in} fG_i(a), j \in X_m^{in} fC_j(a) \right]$$

Multi-person decision-making involves several persons so that the expert knowledge from various persons is utilized to make decisions.

The calculation for this can be given as follows –

Number of persons preferring x_i to $x_j = N(x_i, x_j)$

Total number of decision makers $= n$

Then,

$$SC(x_i, x_j) = \frac{N(x_i, x_j)}{n}$$

3.7.3 Multi-Objective Decision-Making

Many simple decision processes are based on a single objective, such as minimizing cost, maximizing profit, minimizing run time, and so forth. Often, however, decisions must be made in an environment where more than one objective constrains the problem, and the relative value of each of these objectives is different. For example, suppose we are designing a new computer, and we want simultaneously to minimize cost, maximize CPU, maximize random access memory (RAM), and maximize reliability. Moreover, suppose the cost is the most important of our objectives and the other three (CPU, RAM, reliability) carry lesser but equal weight when compared with cost. Two primary issues in multi-objective decision-making are to acquire meaningful information regarding the satisfaction of the objectives by the various choices (alternatives) and to rank or weight the relative importance of each of the objectives. The approach illustrated in this section defines a decision calculus that requires only *ordinal* information on the ranking of preferences and importance weights.

Multi-objective decision-making occurs when there are several objectives to be realized. There are following two issues in this type of decision making

- To acquire proper information related to the satisfaction of the objectives by various alternatives.
- To assign a weight to the relative importance of each objective.

Mathematically we can define a universe of n alternatives as

$$A = [a_1, a_2, \ldots, a_i, \ldots, a_n]$$

And the set of "m" objectives as $O = [o_1, o_2, \ldots, o_i, \ldots, o_n]$

3.7.4 Multi-Attribute Decision-Making

Multi-attribute decision-making takes place when the evaluation of alternatives can be carried out based on several attributes of the object. The attributes can be numerical data, linguistic data, and qualitative data.

Mathematically, the multi-attribute evaluation is carried out on the basis of a linear equation as follows:

$$Y = A_1 X_1 + A_2 X_2 + \ldots + A_i X_i + \ldots + A_r X_r$$

3.7.5 Fuzzy Bayesian Decision-Making

Classical statistical decision-making involves the notion that the uncertainty in the future can be characterized probabilistically, as discussed in the introduction to this chapter.

When we want to make a decision among various alternatives, our choice is predicated on information about the future, which is normally discretized into various 'states of nature.' If we knew with certainty the future states of nature, we would not need an analytic method to assess the likelihood of a given outcome. Unfortunately, we do not know what the future will entail, so we have devised methods to make the best choices given an uncertain environment. Classical Bayesian decision methods presume that future states of nature can be characterized as probability events. For example, consider the condition of 'cloudiness' in tomorrow's weather forecast by discretizing the state space into three levels and assessing each level probabilistically: the chance of a very cloudy day is 0.5, a partly cloudy day is 0.2, and a sunny (no clouds) day is 0.3. By convention the probabilities sum to unity. The problem with the Bayesian scheme here is that the events are vague and ambiguous. How many clouds does it take to transition between very cloudy and cloudy? If there is one small cloud in the sky, does this mean it is not sunny? This is the classic sorites paradox.

The following material first presents Bayesian decision-making and then starts to consider ambiguity in the value of new information, in the states of nature, and in the alternatives in the decision process. Examples will illustrate these points.

First, we shall consider the formation of probabilistic decision analysis. Let $S = \{s_1, s_2, ..., sn\}$ be a set of possible states of nature; and the probabilities that these states will occur are listed in a vector,

$$\mathbf{P} = \{p(s_1), p(s_2), ..., p(s_n)\} \quad \text{where} \quad \sum_{i=1}^{n} p(s_i) = 1$$

The probabilities expressed in the equation above. are called 'prior probabilities' in Bayesian jargon, because they express prior knowledge about the true states of nature. Assume that the decision-maker can choose among m alternatives, $A = \{a_1, a_2, ..., am\}$, and for a given alternative aj we assign a utility value, uji, if the future state of nature turns out to be state si. These utility values should be determined by the decision-maker, since they express value or cost for each alternative–state pair (i.e., for each aj–si combination).

The utility values are usually arranged in a matrix of the form shown in Table 3.1. The expected utility associated with the jth alternative would be

$$E(u_j) = \sum_{i=1}^{n} u_{ji} p(s_i)$$

TABLE 3.1

Utility Matrix

Stats s_i Action a_j	s_1	s_1	...	s_a
a_1	u_{11}	u_{12}	...	u_{1n}
⋮	⋮	⋮		⋮
a_m	u_{m1}	u_{m2}	...	u_{mn}

The most common decision criterion is the *maximum* expected utility among all the alternatives, that is,

$$E\left(u^*\right) = \max_j E\left(u_j\right)$$

which leads to the selection of alternative a_k if $u^* = E\left(u_k\right)$.

3.7.6 Summary

Here, 0 various fuzzy decision-making methods are discussed. The fuzzy Bayesian decision-making is providing both fuzzy and random uncertainty. Multi-objective decision-making was included based on several objectives. The main processes involved in decision-making are the determination of sets of alternatives and evaluating among them. The form of imprecision is important for dealing with many of the uncertainties within human systems.

3.8 Fuzzy Logic Control Systems

3.8.1 Introduction

Control applications are the kinds of problems for which fuzzy logic has had the greatest success and acclaim. Many of the consumer products that we use today involve fuzzy control. And, even though fuzzy control is now a standard in many industries, the teaching of this subject on academic campuses is still far from being a *standard* offering. But a paradigm shift is being realized in the area of fuzzy control, given its successes for some problems where classical control has not been effective or efficient. It was not long ago that fuzzy logic and fuzzy systems were the subject of ridicule and scorn in the scientific communities, but the control community moved quickly in accepting the *new paradigm* and its success is now manifested in the marketplace. For pattern recognition and clustering analysis, please see works Bezdek, 1981 and Chiu, 1994.

Control systems abound in our everyday life. Perhaps we do not see them as such, perhaps because some of them are larger than what a single individual can deal with—but they are ubiquitous. For example, economic systems are large global systems that can be controlled; ecosystems are large, amorphous, and long-term systems that can be controlled. Systems that can be controlled have three key features: inputs, outputs, and control parameters (or actions) which are used to perturb the system into some desirable state. The system is monitored in some fashion and left alone if the desired state is realized, or perturbed with control actions until the desired state is reached (Larsen, 1980; Mendel, 1985; Kosco, 1997).

Usually, the control parameters (actions) are used to perturb the inputs to the system. For example, in the case of economic systems the inputs might be the balance of trade index, the federal budget deficit, and the consumer price index; outputs might be the inflation rate and the Dow Jones Industrial index; a control parameter might be the federal lending rate that gets adjusted occasionally by the US Federal Reserve Board. In the case of ecosystems, the inputs could be the rate of urbanization, automobile traffic, and water use; the outputs could be reductions in green spaces, or habitat erosion; a control action could be federal laws and policy on pollution prevention. Other everyday control situations are evident in our daily lives. Traffic lights are control mechanisms: inputs are arrival rates of cars at an intersection and time of day, outputs are the length of the lines at the lights, and the control parameters are the length of the various light actions (green, yellow, green arrow, etc.). And, construction projects involve control scenarios. The inputs on these projects would include the weather, availability of materials, and labor; outputs could be the daily progress toward goals and the dates of key inspections; the control actions could include rewards for finishing on time or early and penalties for finishing the project late. There are numerous texts which focus just on fuzzy control; a single chapter on this subject could not possibly address all the important topics in this field.

Physical control systems are typically one of two types: open-loop control systems, in which the control action is independent of the physical system output, and closed-loop control systems (also known as *feedback control systems*), in which the control action depends on the physical system output. Examples of open-loop control systems are a toaster, in which the amount of heat is set by a human, and an automatic washing machine, in which the controls for water temperature, spin-cycle time, and so on are preset by the human. In both these cases the control actions are not a function of the output of the toaster or the washing machine.

Examples of feedback control are a room temperature thermostat, which senses room temperature and activates a heating or cooling unit when a certain threshold temperature is reached, and an autopilot mechanism, which makes automatic course corrections to an aircraft when heading or altitude deviations from certain preset values are sensed by the instruments in the plane's cockpit.

In order to control any physical variable, we must first measure it. The system for measurement of the *controlled signal* is called a *sensor*. The physical system under control is called a *plant*. In a closed-loop control system, certain forcing signals of the system (the *inputs*) are determined by the responses of the system (the *outputs*). To obtain satisfactory responses and characteristics for the closed-loop control system, it is necessary to connect an additional system, known as a *compensator*, or a *controller*, into the loop.

Control systems are sometimes divided into two classes. If the object of the control system is to maintain a physical variable at some constant value in the presence of disturbances, the system is called a *regulatory* type of control, or a regulator. Sometimes this type is also referred to as *disturbance-rejection*. The room temperature control and autopilot are examples of regulatory controllers. The second class of control systems are *set-point tracking* controllers. In this scheme of control, a physical variable is required to follow or track some desired time function.

3.8.2 Control System Design

Following are the steps involved in designing fuzzy logic control (FLC) systems:

- **Identification of variables:** The input, output, and state variables of the plant which is under consideration must be identified.
- **Fuzzy subset configuration:** The universe of information is divided into a number of fuzzy subsets and each subset is assigned a linguistic label. Always make sure that these fuzzy subsets include all the elements of the universe.
- **Obtaining a membership function:** Obtain the membership function for each fuzzy subset that we got in the previous step.
- **Fuzzy rule base configuration:** Formulate the fuzzy rule base by assigning the relationship between fuzzy input and output.
- **Fuzzification:** The fuzzification process is initiated in this step.
- **Combining fuzzy outputs:** By applying fuzzy approximate reasoning, locate the fuzzy output and merge them.
- **Defuzzification:** Finally, initiate the defuzzification process to form a crisp output.

3.8.3 Operation of the FLC system

The general problem of the feedback control system design is defined as obtaining a generally nonlinear vector-valued function $\mathbf{h}(\ )$, defined for some time, t, as follows:

$$\mathbf{u}(t) = \mathbf{h}[t, \mathbf{x}(t), \mathbf{r}(t)]$$

where $\mathbf{u}(t)$ is the control input to the plant or process, $\mathbf{r}(t)$ is the system reference (desired) input, and $\mathbf{x}(t)$ is the system state vector; the state vector might contain quantities like the system position, velocity, or acceleration. The feedback control law $\mathbf{h}$ is supposed to stabilize the feedback control system and result in a satisfactory performance.

In the case of a time-invariant system with a regulatory type of controller, where the reference input is a constant setpoint, the vast majority of controllers are based on one of the general models, that is, either full state feedback or output feedback as shown in the following:

$$\mathbf{u}(t) = \mathbf{h}[\mathbf{x}(t)]$$

$$\mathbf{u}(t) = \mathbf{h}\left[y(t), \dot{y}, \int y\mathrm{d}t\right]$$

where $y()$ is the system output or response function. In the case of a simple single-input, single-output system and a regulatory type of controller, the function $\mathbf{h}$ takes one of the following forms:

$$\mathbf{u}(t) = K_P \cdot e(t)$$

for a proportional, or P, controller;

$$\mathbf{u}(t) = K_P \cdot e(t) + K_I \cdot \int e(t)\mathrm{d}t$$

for a proportional-plus-integral, or PI, controller;

$$\mathbf{u}(t) = K_P \cdot e(t) + K_D \cdot \dot{e}(t)$$

for a proportional-plus-derivative, or PD, controller

$$\mathbf{u}(t) = K_P \cdot e(t) + K_1 \cdot \int e(t)\mathrm{d}t + K_D \cdot \dot{e}(t)$$

For a proportional-plus-derivative-plus-integral, or PID, controller, where $e(t)$, times $\dot{e}(t)$, and $\int e(t)\mathrm{d}t$ are the output error, error derivative, and error integral, respectively; and

$$\mathbf{u}(t) = -\left[k_1 \cdot x_1(t) + k_2 \cdot x_2(t) + \cdots + k_n \cdot x_n(t)\right]$$

for a full state-feedback controller.

3.8.4 FLC System Models

An important part of fuzzy reasoning is represented by Fuzzy Logic Control (FLC), derived from control theory based on mathematical models of the open-loop process to be controlled. FLC has been successfully applied to a wide variety of practical problems: control of warm water, robotics, heat exchange, traffic junction, cement kiln, automobile speed, automotive engineering, model car parking and turning, power systems and nuclear reactors, on-line shopping, washing machines, etc.

This points out that fuzzy control has been effectively used in the context of complex ill-defined processes, especially those that can be controlled by a skilled human operator without the knowledge of their underlying dynamics. In this sense, neural and adaptive fuzzy systems have been compared to classical control methods. It has been remarked that they are model-free estimators, that is, they estimate a function without requiring a mathematical description of how the output functionally depends on the input; they simply learn from samples. However, some people criticized fuzzy control because of the very fundamental question 'Why does a fuzzy rule-based system have such good performance for a wide variety of practical problems?' remains unanswered. A first approach to answer this fundamental question in a quantitative way was presented by Wang (1992), where he proved that a particular class of FLC systems are universal approximators: they are capable of approximating any real continuous function on a compact set to arbitrary accuracy. This class is characterized by:

- Gaussian membership functions
- Product fuzzy conjunction
- Product fuzzy implication
- Center of area defuzzification

It is proved that a modification of Sugeno-type fuzzy controllers gives universal approximators. Although both results are very important, many real fuzzy logic controllers do not belong to these classes, because other membership functions are used, other inference mechanisms are applied, or other types of rules are used. The question 'What other types of fuzzy logic controllers are universal approximators?' still remains unanswered. This problem was solved in the way that a large number of classes of FLC systems are also universal approximators. The most popular FLC systems are: Mamdani, Tsukamoto, Sugeno, and Larsen which work with crisp data as inputs.

3.8.5 Summary

In this section, the basic structure and design aspects of fuzzy logic controllers are explained. The fuzzy logic controller is supported by fuzzy control rules which link the input–output relationship of a controlled system. The stability, observability, and controllability are well established in modern control theory. The object evaluation of fuzzy control rules predicts the present and future control actions.

3.9 Merits and Demerits of Fuzzy Logic

3.9.1 Introduction

Every intelligent technique has merits and demerits based on the problem-solving point of view. Proper selection of techniques for the application area is crucial for an optimal solution. Fuzzy logic may be preferable over other techniques in various scenarios related to constraints. Many papers have reported various advantages and disadvantages of fuzzy logic applications (Cox, 1994; Jang, 1991; Bezdek, 1993).

3.9.2 Merits of Fuzzy Logic

- Fuzzy logic describes systems in terms of a combination of numerics and linguistics (symbolics). This has advantages over pure mathematical (numerical) approaches or pure symbolic approaches because very often system knowledge is available in such a combination.
- Problems for which an exact mathematically precise description is lacking or is only available for very restricted conditions can often be tackled by fuzzy logic, provided a fuzzy model is present.
- Fuzzy logic sometimes uses only approximate data, so simple sensors can be used.
- The algorithms can be described with little data, so little memory is required.
- The algorithms are often quite understandable.
- Fuzzy algorithms are often robust, in the sense that they are not very sensitive to changing environments and erroneous or forgotten rules.
- The reasoning process is often simple, compared to computationally precise systems, so computing power is saved. This is a very interesting feature, especially in real-time systems.
- Fuzzy methods usually have a shorter development time than conventional methods.

3.9.3 Demerits of Fuzzy Logic

- Fuzzy logic amounts to function approximation in the case of crisp-input/crisp-output systems. This means that in many cases, using fuzzy logic is just a different way of performing the interpolation.
- In areas that have good mathematical descriptions and solutions, the use of fuzzy logic most often may be sensible when computing power (i.e., time and memory) restrictions are too severe for a complete mathematical implementation.
- Careful analysis of comparison examples 'proving' the superiority of fuzzy logic often shows that they compare the fuzzy approach with a very simple, non-optimized conventional approach.
- Proof of characteristics of fuzzy systems is difficult or impossible in most cases because of lacking mathematical descriptions; this is an especially important research item in the area of stability of control systems.

3.10 Fuzzy Rule-Based or Inference Systems

3.10.1 Introduction

The basic structure of the fuzzy rule-based system involves four principal components: fuzzification interface, where the values of inputs are measured and fuzzified and the input range is mapped into the suitable universe of discourse; knowledge-base, which involves a numeric 'database' section and a fuzzy (linguistic) rule-base section; fuzzy inference mechanism or engine, which constitutes the core of the fuzzy logic control and involves decision-making logic (fuzzy reasoning such as product, max–min composition, etc); and defuzzification interface, which maps the range of output variables into corresponding universe of discourse and defuzzifies the results of the fuzzy inference mechanism.

In a complex system, which is usually the case, the fuzzy rule-based system construction is limited (manipulation and verbalization by expert). Therefore, the possibility of inducing and learning the rules from data can be investigated and implemented successfully. These systems are called fuzzy adaptive systems (FAS). For given relevant variables, the fuzzy rule-based system has to deliver the response close to the observed ones. In other words, on the basis of user-defined input membership functions and input–output sets, FAS can determine the output membership functions and defuzzified outputs.

Fuzzy inference is the process of formulating the mapping from a given input to an output using fuzzy logic. The mapping then provides a basis from which decisions can be made or patterns discerned. Fuzzy inference systems have been successfully applied in fields such as automatic control, data classification, decision analysis, expert systems, and computer vision. Because of its multidisciplinary nature, fuzzy inference systems are associated with a number of names, such as fuzzy-rule-based systems, fuzzy expert systems, fuzzy modeling, fuzzy associative memory, fuzzy logic controllers, and simply (and ambiguously) fuzzy systems. Two types of fuzzy inference systems can be implemented in the Fuzzy Logic Technique (FLT): Mamdani type (Mamdani, 1977) and Takagi–Sugeno type (Sugeno, 1985).

3.10.2 Mamdani Fuzzy Inference System

The Mamdani Fuzzy Inference System (FIS) is the most commonly used fuzzy methodology. This method is the first control system built using fuzzy set theory. Within this inference, the fuzzy sets from the consequent of each rule are combined through the aggregation operator to get the resulting fuzzy set. After the aggregation process, there is a fuzzy set for each output variable that needs defuzzification. Figure 3.6 shows the structure of Mamdani Inference System. Figure 3.7 shows the correlation minimum of action for truncated output surface. Figure 3.8 shows the truncated output

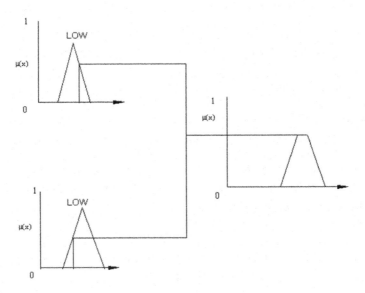

FIGURE 3.7
Correlation minimum of action for truncating output surface.

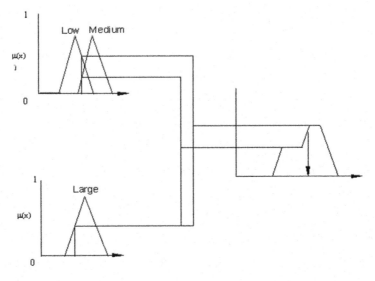

FIGURE 3.8
Truncated output surface for more than one rule is fired.

surface for more than one rule is fired. These figures are very crucial toward understanding the Mamdani inference system. The following steps are followed in the Mamdani FIS:

1. Fuzzification of input and output variable by considering convenient linguistic subsets such as high, medium, heavy, light, hot, warm, big, small, etc.
2. Construction of fuzzy IF–THEN rules based on the expert knowledge and/or on the basis of available literature. The rules related to the combined linguistic subsets within the premise part are combined interchangeably with logical 'AND' or 'OR' conjunction.
3. The implication part of a system is defined as the shape of the consequent based on the premise (antecedent) part.
4. Finally, the result is a fuzzy set and requires defuzzification to arrive at a crisp value, which is required by the analyst or decision-maker.

3.10.3 Takagi–Sugeno (TS) Fuzzy Inference System

The Takagi–Sugeno FIS is similar to the Mamdani method in many respects. The first two parts of the fuzzy inference process (fuzzifying inputs and applying fuzzy operators) are exactly the same. The main difference between the Mamdani FIS and the TS FIS is the output membership functions are linear or constant.

Since the terms used to describe the various parts of the fuzzy inference process are far from standard, we will try to be as clear as possible about the different terms introduced in this section. The process of the fuzzy inference system involves all of the procedures that are described in the following sections.

3.10.4 A Linguistic Variable

A linguistic variable is a variable whose values are the words of sentences in a natural or artificial language; it takes fuzzy variables as its values. For instance, let 'rainfall' be a linguistic variable. It takes the fuzzy variables as its values, such as the terms 'low, medium, high,' etc. Conditional fuzzy statements, often termed 'production rules,' are actually the relations between linguistic variables. Production rules are generally expressed in the pattern of IF–THEN, which consists of a condition (IF part) and conclusion (THEN part). Thus, they are also called fuzzy IF–THEN rules. The IF part can consist of more than one precognition linked together by linguistic conjunctions such as AND and OR.

IF–THEN Rules

Fuzzy sets and fuzzy operators are the subjects and verbs of fuzzy logic. These IF–THEN rule statements are used to formulate the conditional statements that comprise fuzzy logic. A single fuzzy IF–THEN rule assumes the form

**IF X *is High,* Then Y *is High*

Where *high is* linguistic values defined by fuzzy sets on the ranges (universes of discourse) X and Y, respectively. The if part of the rule 'X is high' is called the *antecedent* or premise, while the then part of the rule 'Y is high' is called the *consequent* or conclusion.

3.10.5 Membership Functions

Membership functions are the most subjective part of fuzzy logic modeling. Each variable must have membership functions, usually represented by linguistic terms and defined for the entire range of possible values. The linguistic terms normally describe a concept related to the value of the variable, such as low, medium, and high. These linguistic membership functions define the degree to which a particular numerical value of a variable fits the concept expressed by the linguistic term. The range of values from zero (not part of the set) to one perfectly represents the linguistic concept. Membership functions can take several shapes, depending on the philosophy behind the concept of the linguistic term. Other shapes may be able to reflect natural phenomena but may require complex equations to a model which can drastically increase the size and complexity of the fuzzy system.

A fuzzy singleton is a fuzzy set whose support is a single point in U with a membership function of one. Singletons are easily represented in a computer and allow for simpler defuzzification algorithms. They are, therefore, frequently used to describe fuzzy outputs. Each crisp input into a fuzzy system can have multiple labels assigned to it. In general, the greater the number of labels assigned to describe an input variable, the higher the resolution of the resultant fuzzy control system, resulting in a smoother response.

However, a large number of labels requires added computation time. Moreover, an excessive number of labels can lead to an unstable fuzzy system. As a result, the most common number of labels for each variable in a fuzzy system falls between 3 and 9. The number is usually an odd number—3, 5, 7, 9; although this is not a requirement, it is not uncommon.

In general, one rule by itself doesn't do much good. What is needed are two or more rules that can play off of one another. The output of each rule is a fuzzy set. The output fuzzy sets for each rule are then *aggregated* into a single output fuzzy set. Finally, the resulting set is *defuzzified* or resolved to a single number. Fuzzy inference systems show how the whole process works from beginning to end for a particular type of fuzzy inference system called a Mamdani type.

Fuzzy Logic has five steps: fuzzification of the input variables, application of the fuzzy operator (AND or OR) in the antecedent, implication from the antecedent to the consequent, aggregation of the consequents across the rules, and defuzzification. These odd and sometimes cryptic terms have very specific meanings that we'll define carefully as we step through each of them in more detail.

1. **Fuzzify the inputs:** Fuzzification is the first step in the computation of a fuzzy system and it must be performed for each input variable. It is a process of mapping the crisp numbers into fuzzy domains using the membership functions of linguistic variables to compute each term's degree of validity at a specific operation point of the process. The result of fuzzification is used as an input for the fuzzy inference engine.

2. **Apply fuzzy operator or fuzzy rule inference:** Once the inputs have been fuzzified, we know the degree to which each part of the antecedent has been satisfied for each rule. If the antecedent of a given rule has more than one part, the fuzzy operator is applied to obtain one number that represents the result of the antecedent for that rule. This number will then be applied to the output function. The input to the fuzzy operator is two or more membership values from fuzzified input variables. The output is a single truth value.

   The fuzzy rule inference consists of two steps:

   a) Inference, which determines the fuzzy subset of each output variable for each rule. Usually, only MIN or PRODUCT is used as inference rules. In MIN inference, the output membership function is clipped off at a height corresponding to the rule premise's computed degree of truth (fuzzy logic AND). In PRODUCT inference, the output membership function is scaled by the rule premises' computed degree of truth.

   b) Composition, which combines the fuzzy subsets for each output variable einto a single fuzzy subset. Usually, MAX or SUM are used. In MAX composition, the combined output fuzzy subset is constructed by taking the point-wise maximum over all of the fuzzy subsets assigned to variables by the inference.

3. **Apply the implication method:** Before applying the implication method, we must take care of the rule's weight. Every rule has a *weight* (a number between 0 and 1), which is applied to the number given by the antecedent. Generally, this weight is 1 (as it is for this example) and so it has no effect at all on the implication process. From time to time you may want to weight one rule relative to the others by changing its weight value to something other than 1.

   Once proper weighting has been assigned to each rule, the implication method is implemented. A consequent is a fuzzy set represented by a membership function, which appropriately weights the linguistic characteristics that are attributed to it. The consequent is reshaped using a function associated with the antecedent (a single number). The input for the implication process is a single number given by the antecedent, and the output is a fuzzy set. The implication is implemented for each rule.

4. **Aggregate the consequents:** Since decisions are based on the testing of all of the rules in an FIS, the rules must be combined in some manner in order to make a decision. Aggregation is the process by which the fuzzy sets that represent the outputs of each rule are combined into a single fuzzy set. Aggregation only occurs once for each output variable, just prior to the fifth and final step, defuzzification. The input of the aggregation process is the list of truncated output functions returned by the implication process for each rule. The output of the aggregation process is one fuzzy set for each output variable. Figure 3.4 shows the aggregation for inputs

Notice that as long as the aggregation method is commutative (which it always should be), then the order in which the rules are executed is unimportant.

5. **Defuzzification:** The output of a fuzzy algorithm is a fuzzy subset, which is not the crisp number required. The process of retranslating a fuzzy output into a crisp value is termed defuzzification, in which a defuzzification algorithm (defuzzifier) selects the best crisp value to be the output of the fuzzy system. As much as fuzziness helps the rule evaluation during the intermediate steps, the final desired output for each variable is generally a single number. However, the aggregate of a fuzzy set encompasses a range of output values and so must be defuzzified in order to resolve a single output value from the set. The diagrammatic representation of defuzzification is shown in Figure 3.5. Defuzzification can be performed using several different methods. The following defuzzification methods are of practical importance:

   a) **Center of area (CoA):** The CoA method is often referred to as the Center of Gravity method because it computes the centroid of the composite area representing the output fuzzy term. This method is one of the most common of the defuzzification methods and consists of selecting the value corresponding to the center of gravity for the solution set.

   b) **Center of maximum (CoM):** In the CoM method, only the peaks of the membership functions are used. The defuzzified crisp compromise value is determined by finding the place where the weights are balanced. Thus, the areas of the membership functions play no role and only the maxima (singleton memberships) are used. The crisp output is computed as a weighted mean of the term membership maxima weighted by the inference results.

c) **Mean of maximum (MoM):** The MoM is used only in some cases where the CoM approach does not work. In this maximum inference value is considered for computing crisp output.

d) **Mean of minima (MoM):** This method of defuzzification is the same as Mean of Maxima, but for defuzzification, minimum inference value is considered for computing crisp output.

3.10.6 Strategy of Fuzzy Logic Systems

An FLS receives a crisp input and may deliver either a fuzzy set or a crisp value. The basic FLS contains four components: a rule set, a fuzzifier, an inference engine, and a defuzzifier. Rules may be provided by experts or can be extracted from numerical data. In either case, the engineering rules are expressed as a collection of IF–THEN statements. These statements are related to fuzzy sets associated with linguistic variables.

The fuzzifier maps the input crisp numbers into the fuzzy sets to obtain degrees of membership. It is needed in order to activate rules, which are in terms of the linguistic variables. The inference engine of the FLS maps the antecedent fuzzy (IF part) sets into consequent fuzzy sets (THEN part). This engine handles the way in which the rules are combined. In practice, only a very small number of rules are actually used in engineering applications of fuzzy logic (Wang, 1994; Wang and Mendel, 1991).

In most applications, crisp numbers must be obtained at the output of an FLS. The defuzzifier maps output fuzzy sets into a crisp number, which becomes the output of the FLS. In a control application, for example, such a number corresponds to a control action.

In fuzzy logic there are five steps of the fuzzy logic system:

1. Fuzzification of input variables
2. Apply fuzzy operator (AND or OR)
3. Apply implication from the antecedent to the consequent
4. Aggregation of all output
5. Defuzzification

A fuzzy inference system operates by repeating a cycle of the following steps. First, measurements are taken of all variables that represent relevant conditions of the controlled process. Next, these measurements are converted into appropriate fuzzy sets to express measurement uncertainties. This step is called a fuzzification. The fuzzified measurements are then used by the inference engine to evaluate the control rules stored in the fuzzy rule base. The result of this evaluation is a fuzzy set defined on the universe of possible actions. This fuzzy set is then converted, in the final step of the cycle, into a single (crisp) value that, in some sense, is the best representative of the fuzzy set. This conversion is called a defuzzification.

The defuzzified values represent actions taken by the fuzzy controller in individual control cycles.

3.10.7 Summary

In this section, the fuzzy rule-based system methodology has been discussed. Various inference systems, such as the Mamdani and Takaki–Sugeno methods, are briefly explained. Also, various defuzzification methods are discussed. Mathematical functions of various membership functions have been discussed. The comparison between Mamdani and Sugeno has been discussed.

References

Bezdek, J.C. 1981. *Pattern Recognition with Fuzzy Objective Function Algorithms*, Plenum Press, New York, 1981.

Bezdek, J.C. 1993. Fuzzy models—what are they and why—editorial, *IEEE Transactions on Fuzzy Systems*, 1, 1–5.

Chiu, S. 1994. Fuzzy model identification based on cluster estimation, *Journal of Intelligent & Fuzzy Systems*, 2(3), 267–278.

Cox, E. 1994. *The Fuzzy Systems Handbook*, Cambridge, MA: Academic Press.

Dubois, D. and Prade, H. 1980. *Fuzzy Sets and Systems: Theory and Applications*, Academic Press, New York.

Giarratano, J. and Riley, G. 1993. *Expert Systems: Principles and Programming*, PWS-Kent, Boston, MA.

Jang, J.-S.R. 1991. Fuzzy modeling using generalized neural networks and Kalman filter algorithm, Proceedings of the Ninth National Conference on Artificial Intelligence (AAAI-91), pp. 762–767.

Kaufmann, A. and Gupta, M.M. 1985. *Introduction to Fuzzy Arithmetic*, Van Nostrand, New York.

Kim, Y.M. and Mendel, J.M. 1995. Fuzzy basis functions: comparison with other basis functions, USC SIPI Report, no. 299.

Kosko, B. 1992. Fuzzy systems as universal approximators, Proceedings of the IEEE International Conference on Fuzzy Systems. San Diego, CA, pp. 1151–1162.

Kosko, B. 1994. Fuzzy systems as universal approximators, *IEEE Transactions on Computers*, 43(11), 1329–1333.

Kosko, B. 1997. *Fuzzy Engineering*, Prentice Hall, Upper Saddle River, NJ.

Larsen, P.M. 1980. Industrial applications of fuzzy logic control, *International Journal of Man-Machine Studies*, 12(1), 3–10.

Mamdani, E.H. 1977. Applications of fuzzy logic to approximate reasoning using linguistic synthesis, *IEEE Transactions on Computers*, 26(12), 1182–1191.

Mendel, J.M. 1995. Fuzzy logic systems for engineering: a tutorial, *Proceedings of the IEEE*, 83(3), 345–377.

Schweizer, B. and Sklar, A. 1963. Associative functions and abstract semi-groups, *Publicationes Mathematicae Debrecen*, 10, 69–81.

Wang, L.-X. 1994. *Adaptive Fuzzy Systems and Control: Design and Stability Analysis*, Prentice Hall.

Wang, L.-X. and Mendel, J.M. 1991. Generating fuzzy rules by learning from examples, Proceedings of the IEEE International Symposium on Intelligent Control, Arlington, VA, pp. 263–268.

Zadeh, L.A. 1965. Fuzzy sets. *Information and Control*, 8, 338–353.

4

Support Vector Machine

4.1 Introduction to Statistical Learning Theory

Statistical learning refers to a vast set of mathematical implementations for understanding data. These implementations can be classified as supervised or unsupervised. Broadly speaking, supervised statistical learning involves building a statistical model for predicting, or estimating, an output based on one or more inputs. Problems of this nature occur in fields as diverse as business, medicine, astrophysics, and public policy. With unsupervised statistical learning, there are inputs but no supervising output; nevertheless, we can learn relationships and structure from such data. The main goal of statistical learning theory is to provide a framework for studying the problem of inference (i.e., gaining knowledge, making predictions, making decisions and constructing models from a set of data). This is studied in a statistical framework; that is, there are assumptions of a statistical nature about the underlying phenomena (in the way the data is generated). Indeed, a theory of inference should be able to give a formal definition of words like learning, generalization, and overfitting, as well as to characterize the performance of learning algorithms so that, ultimately, it may help design better learning algorithms. There are two goals: make things more precise and derive new or improved algorithms.

The support vector machine (SVM) has been introduced to various pattern classification and function approximation problems. Pattern classification is used to classify some objects into one given category called classes. The classifier, which is computer software, is developed in such a way that the objects are classified correctly with acceptable accuracy. The inputs given to the classifier are called features and they are representing each class well or the data belongs to different classes are well separated in the input space.

4.2 Support Vector Classification

4.2.1 Hard Margin SVM

Hard margin SVM can work only when data is completely linearly separable without any errors (noise or outliers). In case of errors, either the margin is smaller or the hard margin SVM fails. Soft margin SVM was proposed by Vapnik to solve this problem by introducing slack variables. The SVM technique is a classifier that finds a hyperplane or a function that correctly separates two classes with a maximum margin. Another method for increasing the robustness of SVM is to use the hard margin loss, where the number of misclassifications is minimized. The computational complexity of using the hard margin loss has often been used as justification for a continuous measure of error. One can formulate discrete SVM (DSVM) that uses the hard margin loss for SVM with a linear kernel and linearized margin term, using heuristics for solving instances that do not guarantee global optimality. Few researchers have extended their formulation and technique to soft margin SVM and to fuzzy SVM. Brooks (2011) approximated the hard margin loss for SVM with continuous functions and used an iterative reweighted least-squares method for solving instances that also does not guarantee global optimality. Figure 4.1 shows linear classifiers (hyperplane) in two-dimensional spaces.

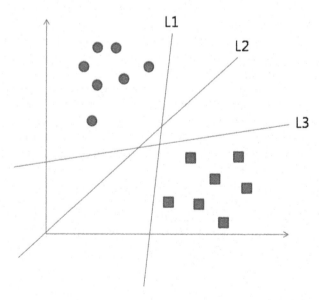

FIGURE 4.1
Linear classifiers (hyperplane) in two-dimensional spaces.

4.2.2 Soft Margin SVM

The hard-maximal margin classifier is an important concept, but it has two problems. First, the hard margin classifier can be very brittle, since any labeling mistake on support vectors will result in a significant change in the decision hyperplane. Second, training data is not always linearly separable, and when it is not, we are forced to use a more powerful kernel which might result in over-fitting. To be able to tolerate noise and outliers, we need to take into consideration the positions of more training samples than just those closest to the boundary. This is done generally by introducing slack variables and a soft margin classifier.

4.2.3 Mapping to High-Dimensional Space

4.2.3.1 Kernel Tricks

One of the crucial ingredients of SVMs is the so-called kernel trick for the computation of dot products in high-dimensional feature spaces using simple functions defined on pairs of input patterns. This trick allows the formulation of nonlinear variants of any algorithm that can be cast in terms of dot products, SVMs being but the most prominent example. Although the mathematical result underlying the kernel trick is almost a century old, it was only much later that it was made fruitful for the machine learning community. Kernel methods have since led to interesting generalizations of learning algorithms and to successful real-world applications (Schölkopf, 2001). Figure 4.2 shows separation of class.

Kernel Functions: The idea of the kernel function is to enable operations to be performed in the input space rather than the potentially high-dimensional feature space. Hence the inner product does not need to be evaluated in the feature space. We want the function to perform mapping of the

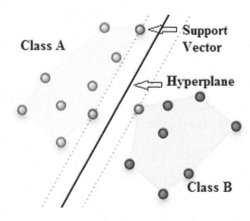

FIGURE 4.2
Separation of class.

attributes of the input space to the feature space. The kernel function plays a critical role in SVM and its performance. It is based upon reproducing kernel Hilbert spaces.

$$K(X,X') = \{\varphi(X)\varphi(X'), \tag{4.1}$$

$$K(X,X') = \sum_{m}^{a} a_m \varphi_m(x) \varphi_m(x'), a_m \geq 0 \tag{4.2}$$

$$\iint K(X,X')g(X)g(X')dxdx' > 0, \ g \varepsilon L_2 \tag{4.3}$$

If K is a symmetric positive definite function, which satisfies Mercer's Conditions, then the kernel represents a legitimate inner product in feature space. The training set is not linearly separable in an input space. The training set is linearly separable in the feature space. This is called the 'Kernel trick.'

Kernels: The significant advantages of SVM is that the generalization performance can be improved by selecting proper kernels. Hence, the selection of kernels for specific applications is crucial.

Linear kernels: If the classification problem is linearly separable in the input space, then mapping the input space into high-dimensional space is not required.

$$K(X,X') = X^T X' \tag{4.4}$$

Polynomial kernel: The polynomial kernel with degree d, where d is a natural number, is given by

$$K(X,X') = \left(X^T X' + 1\right)^d \tag{4.5}$$

Here, 1 is added so that cross terms with degrees equal to or less than d are all included.

When $d = 1$, the kernel is the linear kernel plus 1. Thus, by adjusting b in the decision function, it is equivalent to the linear kernel.

1. **Polynomial:** A polynomial mapping is a popular method for non-linear modeling. The second kernel is usually preferable as it avoids problems with the hessian becoming zero.

$$K(X,X') = \langle X, X' \rangle^d \tag{4.6}$$

$$K(X,X') = [\langle X, X' + 1 \rangle]^d \tag{4.7}$$

2. **Gaussian Radial Basis Function:** Radial basis functions most commonly with a Gaussian form.

$$K(X, X') = \exp\left(-\frac{\|X - X'\|^2}{2\sigma^2}\right)$$ (4.8)

3. **Exponential Radial Basis Function:** A radial basis function (RBF) produces a piecewise linear solution which can be attractive when discontinuities are acceptable.

$$K(X, X') = \exp\left(-\frac{\|X - X'\|}{2\sigma^2}\right)$$ (4.9)

4. **Multi-Layer Perceptron:** The long-established multi-layered perceptron (MLP), with a single hidden layer, also has a valid kernel representation.

$$K(X, X') = \tanh\left(\left(\rho(X, X')\right) + \ell\right)$$ (4.10)

There are many more including Fourier, splines, B-splines, additive kernels, and tensor products. If you want to read more on kernel functions you could refer to any text book on SVM. Figure 4.3 shows the Kernel Trick.

4.2.3.2 Normalizing Kernels

Every input variable has a different physical meaning and thus has a different range. For original input range and without the use of appropriate kernels, the result may not be optimal because SVM are not invariant for the linear transformation of inputs. Thus, to make every input variable work equally in classification, normalization of input variables either by scaling the range or by whitening is required.

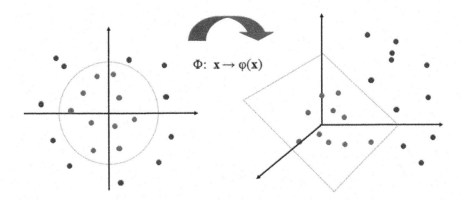

$\Phi: \mathbf{x} \to \varphi(\mathbf{x})$

FIGURE 4.3
Kernel trick.

If we normalize the input variables where the number is very large, the value of the kernel becomes very small or large making the training of SVM difficult. To overcome this, normalizing kernels is highly recommended.

4.2.4 Properties of Mapping Functions Associated with Kernels

To know the neighborhood relations between mapped training data, we have to calculate the Euclidian distances in the input space instead of the feature space. Also, the neighbors nearest to k in the feature space are equivalent to those in the input space.

4.2.5 Summary

Preserving neighborhood relations, one-to-one mapping, nonconvexity of mapped regions, implicit bias terms, and empirical feature space are some of the components attached with mapping functions associated with kernels.

4.3 Multi-Class SVM

4.3.1 Introduction

Support vector machines using direct decision functions, an extension of multi-class problems, is not straightforward. Followings are different types of SVM that handle multi-class problems.

1. One against all SVM
2. Pairwise SVM
3. Error correcting output code SVM
4. All at once SVM

In one against all SVM, an n class problem is converted into two n class problems. For the ith two-class problems, class i is separated from the remaining classes.

4.3.2 Conventional SVM

The SVM has been applied to the problem of predictions and proven to be superior to competing methods, such as the neural network, the linear multiple discriminant approaches, and logistic regression. However, the conventional SVM employs the structural risk minimization principle, thus empirical risk of misclassification may be high, especially when a point to be classified is close to the hyperplane.

4.3.3 Decision Tree–Based SVM

Decision tree–based SVM, which combines support vector machines and decision tree, is an effective way for solving multi-class problems. A problem with this method is that the division of the feature space depends on the structure of a decision tree, and the structure of the tree relates closely to the performance of the classifier. To maintain high generalization ability, the most separable classes should be separated at the upper nodes of a decision tree. Distance is often used as a separability measure between classes, but the distance between class centers cannot reflect the distribution of the classes. After analyzing the tree structure and the classification performance of the decision tree–based SVM, a new separability measure is defined based on the distribution of the training samples in the feature space. The defined separability measure was used in the formation of the decision tree, and an improved algorithm for decision tree–based support vector machine is developed.

Decision tree–based SVM which combines support vector machines and decision tree can be an effective way for solving multi-class problems. This method can decrease the training and testing time, thereby increasing the efficiency of the system. The different ways to construct the binary trees divides the data set into two subsets from root to the leaf until every subset consists of only one class. The construction order of binary tree has great influence on the classification performance. Decision tree–based SVM is also a good way for solving multi-class problems. It combines the SVM and the decision tree approaches for preparing decision-making models. Integrating different models gives better performance than the individual learning or decision-making models. Integration reduces the limitations of the individual model.

4.3.4 Pairwise SVM

To extend binary classifiers to multi-class classification, several modifications have been suggested. A more recent approach used in the field of multiclass and binary classification is pairwise classification. Pairwise classification relies on two input examples, instead of one, and predicts whether the two input examples belong to the same class or to different classes. This is of particular advantage only if a subset of classes is known for training. For later use, a support vector machine that is able to handle pairwise classification tasks is called pairwise SVM. A natural requirement for a pairwise classifier is that the order of the two input examples should not influence the classification result (symmetry). A common approach to enforce this symmetry is the use of selected kernels. For pairwise SVMs, another approach has been

suggested. Some propose the use of training sets with a symmetric structure. We will discuss both approaches to obtain symmetry in a general way.
Based on this, we will provide conditions when these approaches lead to the
same classifier. Moreover, we show empirically that the approach of using
selected kernels is three to four times faster in training. A typical pairwise
classification task arises in face recognition. There, one is often interested in
the interclass generalization, where none of the persons in the training set is
part of the test set. We will demonstrate that training sets with many classes
(persons) are needed to obtain a good performance in the interclass generalization. The training on such sets is computationally expensive. Therefore,
we discuss an efficient implementation of pairwise SVMs. This enables the
training of pairwise SVMs with several millions of pairs. In this way, for
the labeled faces in the wild database, a performance is achieved which is
superior to the current state of the art (Brunner et al., 2012). Figure 4.4 shows
the SVM classification function: the hyperplane maximizing the margin in a
two-dimensional space.

4.3.5 Summary

Here, various classes of SVM have been discussed. Conventional SVM, decision tree–based SVM, and pairwise SVM are the classes used as per their
suitability. Pairwise classification relies on two input examples, instead of
one, and predicts whether the two input examples belong to the same class

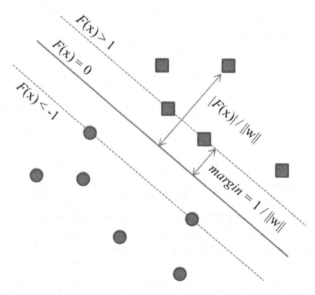

FIGURE 4.4
SVM classification function: the hyperplane maximizing the margin in a two-dimensional
space

or to different classes. To maintain high generalization ability, the most separable classes should be separated at the upper nodes of a decision tree.

4.4 Various SVMs

4.4.1 Introduction

Various SVMs are introduced in the following sections: least square, linear programming, sparse, robust, and Bayesian SVMs are discussed briefly. They are used as per problem suitability.

4.4.2 Least Square SVM

A modified version of SVM, called least square SVM (LSSVM), only considers equality constraints instead of inequalities. In order to perform the decision-making for the extracted features from the wavelet-based technique, the LSSVM technique is employed here as a classifier. Due to the equality constraints in the formulation, LSSVM solves a set of linear equations in the dual space instead of solving a quadratic programming problem as would the standard SVM. This simplifies the computation and enhances the speed considerably. The performance of the SVM mostly depends on kernel function and adjustable weight vectors. However, no such method exists which allows one to decide on an appropriate kernel function in a data-dependent way. The RBF kernel is chosen empirically in this application. The proposed technique relies on the basic idea that in order to improve the performance of the LSSVM, the pattern separability or margin between the clusters needs to be increased. Our aim is to update the adjustable weight vectors at the training phase such that all the data points fall outside the region of separation and to enlarge the width of the separable region (Ari et al., 2010).

4.4.3 Linear Programming SVM

Based on the analysis of the conclusions in the statistical learning theory, especially the **Vapnik–Chervonenkis** (VC) dimension of linear functions, linear programming support vector machines are presented including linear programming, linear, and nonlinear SVMs. In linear programming SVMs, in order to improve the speed of the training time, the bound of the VC dimension is loosened properly. Simulation results for both artificial and real data show that the generalization performance of our method is a good approximation of SVMs and the computation complex is largely reduced by our method.

Support vector machine soft margin classifiers are important learning algorithms for classification problems. They can be stated as convex optimization problems and are suitable for a large data set. The linear programming SVM classifier is especially efficient for very large size samples. But little is known about its convergence, compared with the well understood quadratic programming SVM classifier. The classical SVM model, the so-called 1–norm soft margin SVM, was introduced with polynomial kernels and with general kernels by Cortes and Vapnik (1995). Since then, many different forms of SVM algorithms have been introduced for different purposes. Among them the linear programming SVM (LPSVM) is an important one because of its linearity and flexibility for large data settings. The term 'linear programming' means the algorithm is based on linear programming optimization. Correspondingly, the 1–norm soft margin SVM is called quadratic programming SVM (QPSVM), since it is based on quadratic programming optimization (Vapnik, 1995). Many experiments demonstrate that LPSVM is efficient and performs even better than QPSVM for some purposes. For instance, it is capable of solving huge sample size problems, improving the computational speed, and reducing the number of support vectors (Wu and Zhou, 2005).

4.4.4 Sparse SVM

A sparse representation of support vector machines (SVMs) with respect to input features is desirable for many applications. In many machine learning applications, there is a great desire for sparsity with respect to input features. Several factors account for this. First, many real data sets, such as texts and microarray data, are represented as very high-dimensional vectors resulting in great challenges for further processing. Second, most features in high-dimensional vectors are usually non-informative or noisy and may seriously affect the generalization performance. Third, a sparse classifier can lead to a simplified decision rule for faster prediction in large-scale problems.

4.4.5 Robust SVM

The robust SVM aims to solve the overfitting problem with outliers that make the two classes non-separable. Recently, visual tracking has attracted much research attention. It remains a challenging problem due to issues such as complicated appearance and illumination change, occlusion, cluttered background, etc. To build a robust tracker, a variety of appearance models using different learning techniques have been proposed in the literature. According to the learning techniques, these appearance models may be roughly classified into two categories: generative learning-based and discriminative learning-based appearance mode. Generative learning-based appearance models (GLMs) mainly concentrate on how to construct robust object representation in specified feature spaces, including the integral histogram, kernel density estimation, a spatial-color mixture of Gaussian subspace learning, sparse representation, visual tracking decomposition, and so on. A drawback of these methods is that they often ignore the influence of background and consequently suffer from distractions caused by the background regions with similar appearance to foreground objects.

4.4.6 Bayesian SVM

Another important subspace method is the Bayesian algorithm using probabilistic subspace. Different from other subspace techniques, which classify the test face image into M classes of M individuals, the Bayesian algorithm casts the face recognition problem into a binary pattern classification problem with each of the two classes, intrapersonal variation and extra-personal variation, modeled by a Gaussian distribution. After subspace features are computed, most methods use simple Euclidian distance of the subspace features to classify the face images. Recently, more sophisticated classifiers, such as support vector machines, have been shown to be able to further improve the classification performance of the principal component analysis (PCA) and linear discriminant analysis (LDA) subspace features. Given any two classes of vectors, the aim of SVMs is to find one hyperplane to separate the two classes of vectors so that the distance from the hyperplane to the closest vectors of both classes is the maximum. The hyperplane is known as the optimal separating hyperplane. Support vector machines excel at two-class recognition problems and outperform many other linear and nonlinear classifiers. Figure 4.5 shows graphical relationships among α_i, ξ_i, and C.

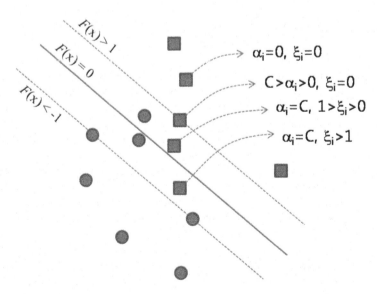

FIGURE 4.5
Graphical relationships among α_i, ξ_i, and C.

4.4.7 Summary

Various SVMs have been discussed in this section: least square, linear programming, sparse, robust, and Bayesian. Given any two classes of vectors, the aim of support vector machines is to find one hyperplane to separate the two classes of vectors so that the distance from the hyperplane to the closest vectors of both classes is the maximum.

4.5 Kernel-Based Methods

4.5.1 Introduction

Conventional pattern classification techniques are extended to incorporate maximizing margins and mapping to a feature space due to the success of SVMs in improving generalization and classification abilities.

4.5.2 Kernel Least Squares

Least square methods in the input space can be readily extended to the feature space using kernel techniques.

4.5.3 Kernel Principal Component Analysis

Principal component analysis (PCA) provides an efficient way to find these underlying gene components and reduce the input dimensions. This linear transformation has been widely used in gene expression data analysis and compression. If the data are concentrated in a linear subspace, PCA provides a way to compress data and simplify the representation without losing much information. However, if the data are concentrated in a nonlinear subspace, PCA will fail to work well. In this case, one may need to consider kernel principal component analysis (KPCA). KPCA is a nonlinear version of PCA. It has been studied intensively in the last several years in the field of machine learning and has claimed success in many applications. In classification and other data analytic tasks it is often necessary to utilize preprocessing on the data before applying the algorithm at hand and it is common to first extract features suitable for the task to solve. Feature extraction for classification differs significantly from feature extraction for describing data. For example, PCA finds directions which have minimal reconstruction error by describing as much variance of the data as possible with m orthogonal directions. Considering the first directions, they need not (and in practice often will not) reveal the class structure that we need for proper classification. The discriminant analysis addresses the following question: Given a data set with two classes, say, which is the best feature or feature set (either linear or nonlinear) to discriminate the two classes? Classical approaches tackle this question by starting with the (theoretically) optimal Bayes classifier and by assuming normal distributions for the classes. Standard algorithms like quadratic or linear discriminant analysis, among them the famous Fisher discriminant, can be derived. Of course, any other model different from a Gaussian for the class distributions could be assumed; this, however, often sacrifices the simple closed-form solution.

4.5.4 Kernel Discriminate Analysis

Linear discriminant analysis (LDA) is a traditional statistical method which has proven successful in classification problems. The procedure is based on an eigenvalue resolution and gives an exact solution of the maximum of the inertia. But this method fails for a nonlinear problem. There have been many works that reported the generalized LDA to nonlinear problems, and an effort is being made to develop a generalized discriminant analysis (GDA) by mapping the input space into a high-dimensional feature space with linear properties. In the new space, one can solve the problem in a classical way, such as the LDA method. The main idea is to map the input space into a convenient feature space in which variables are nonlinearly related to the input space. This fact has been used in some algorithms such as unsupervised learning algorithms. LDA is a standard tool for classification. It is based on a transformation of the input space into a new one. The data are described as a linear combination of the new coordinate values which are called principal components and represent the discriminant axis. As such, for the LDA, the purpose of the GDA method is to maximize the interclass inertia and minimize the intra-class inertia.

4.5.5 Summary

In this section, we have briefly discussed kernel discriminate analysis, kernel least square, and kernel PCA. Kernel PCA is a nonlinear version of PCA. It has been studied intensively in the last several years in the field of machine learning and has claimed success in many applications. The discriminate analysis may be linear and generalized, which are standard tools for classification. It is based on a transformation of the input space into a new one.

4.6 Feature Selection and Extraction

4.6.1 Introduction

This section deals with how unrelated features lead to noise, heavy computation, etc. Also, how the Interdependent Features breaks into different redundant features and targeted to Better Model are discussed.

Feature selection methods are:

- **Filter method:** Applies a statistical measure to assign a scoring to each feature. For example, the Chi-squared test, information gain, and correlation coefficient scores.
- **Wrapper method:** Considers the selection of a set of features as a search problem.
- **Embedded method:** Learns which features best contribute to the accuracy of the model, while the model is being created (e.g., LASSO, Elastic Net, and Ridge Regression).

Controlling the class boundaries is often out of the reach of conventional classifiers, as they don't have the mechanisms to deal with it. Hence, if the number of features such as input variables are large compared to training data, the overlapping of class boundaries may not happen. In this type of scenario, the generalization capability of the conventional classifiers may not be effective. To improve the generalization capability, generating a small set of features from the original input variables by either feature selection or feature extraction is suggested. As SVMs determine the class boundaries straightaway by training, the generalization capability does not decline drastically even though the input variables are large. In the following section, various feature selection methods using SVM will be analyzed. How feature selection affects the generalization capability of an SVM and feature extraction by various methods will also be discussed.

4.6.2 Initial Set of Features

The set of features used is the most influencing factor in realizing a classifier with high generalization capability. As there is no fixed rule in determining an initial set of features for a given classification problem, the only way to determine the set of initial features is through trial and error.

If the number of features is very large and each feature has low classification power, it is better to transform linearly or nonlinearly to produce a reduced set of features. For high classification power, reduction of the set can be achieved by feature selection or feature extraction. Feature selection is reducing redundant or meaningless features so that higher generalization performance and faster classification can be realized.

4.6.3 Procedure for Feature Selection

The objective of feature selection is to select a minimum subset of features from the original set of features which will provide maximum generalization capability. During the process of feature selection, the generalization capability of a subset of features needs to be estimated. This is called the wrapper method. But it is a time-consuming method and so other methods need to be explored, like the filter method linked with the development of various selection criteria.

There are forward or backward selection methods widely used in the selection of criteria. In backward selection, features are tested and deleted one at a time, depending on which causes the least deterioration. This deleting procedure continues until the selection criteria reach a specified value.

In forward selection, features are added to an empty set one at a time, based on which improves the selection criteria the most. This procedure continues until the selection criteria reaches a specified value. As forward or backward selection is low, the addition or deletion of more than one feature at a time based on feature ranking may be possible. Also, the combination of backward and forward selection is possible.

As these selection methods are local optimization techniques, global optimality of feature selection is not confirmed. By the introduction of SVM, various selection methods suitable for SVM have been developed. In most cases, linear SVM is used. If some elements of the coefficient vector of the hyperplane are zero, the deletion of the associated input variables does not change the optimal hyperplane for the remaining variables. The optimal solution changes if variables associated with non-zero elements are deleted. Hence, the magnitude of the margin decreases. In addition to the wrapper and filter methods, the embedded methods combine training and feature selection. Here, the training of SVM results in solving a quadratic optimization problem and feature selection can be done by modifying the objective function. For linear programming, such as SVM with linear kernels, the variables associated with zero coefficients of the separating hyperplane are redundant.

4.6.4 Feature Extraction

For feature extraction, principal component analysis is widely used. Kernel PCA is gaining wider acceptance. It has been shown that the combination of KPCA and linear SVM produces better generalization ability than the nonlinear SVM. KPCA combined with least squares is also used for feature extraction. As PCA does not use the class information, the first principal component is not necessarily useful for class separation. Also, the linear discriminant analysis used for a two-class problem. But application of this is limited to cases where each class consists of one cluster and they are not heavily overlapped. By optimal selection of kernels and their parametric values, kernel discriminated analysis solves the limitation of linear discriminant analysis. It is extended to multi-class problems. For feature selection, kernel discriminant analysis is used as criteria along with kernel selection and feature extraction.

4.6.5 Clustering

SVM can be formulated for one-class problems. It is called one-class classification and is applied to clustering and detection of outliers for both pattern classification and function approximation. Conventional clustering methods, like k-means clustering algorithm and fuzzy-cthe means clustering algorithms, can be extended to the feature space. The domain description defines the region of data by a hyperplane in the feature space. The hyperplane in the feature space corresponds to clustered regions in the input space. Hence, the domain description can be used for clustering. If there are no outliers, that means all the data are in or on the hyperplane; then the problem will be to determine the clusters in the input space.

4.6.6 Summary

Here, feature selection and extraction are discussed. Initializing, selection, and clustering for feature extraction is analyzed. Conventional clustering methods, like k-means clustering algorithm and fuzzy-c means clustering algorithms, can be extended to the feature space. The domain description defines the region of data by a hyperplane in the feature space. For feature extraction, principal component analysis is widely used.

4.7 Function Approximation

4.7.1 Introduction

Support vector regression (SVR), an extension of SVM, has been found to have high generalization capability for various function approximation and time series prediction problems. Extensions of SVM are available for pattern classification and function approximations, such as LP- SVR, v- SVR, LS-SVR.

4.7.2 Optimal Hyperplanes

Development of input-output relationships using input–output pairs is the objective of function approximation. In SVR, mapping the input space into a high-dimensional feature space takes place. In the feature space, optimal hyperplane is to be determined.

4.7.3 Soft Margin Support Vector Regression

Recall that for linear SVM, we are to determine a maximum margin hyperplane $W \cdot X + h = 0$ with the following optimization:

minimize $\dfrac{\|W\|^2}{2}$

subject to $y_i : (W:x_i + b) \geq 1; i = 1; 2; :::; n$

In soft margin SVM, we consider the similar optimization technique except with a relaxation of inequalities, so that it also satisfies the case of linearly not separable data.

To do this, soft margin SVM introduces slack variable (ξ), a positive value, into the constraint of an optimization problem.

Thus, for soft margin we rewrite the optimization problem as follows:

minimize $\dfrac{\|W\|^2}{2}$

subject to $(W:xi + b) \geq 1 - \xi_i$; if $y_i = +1$
$(W:xi + b) \leq -1 + \xi i$; if $y_i = -1$
where V_i ; $\xi_i \geq 0$.

Thus, in soft margin SVM, we are to calculate W, b, and ξ_i is a solution to learn SVM.

4.7.4 Model Selection

In SVM, the model selection means selection of optimal parameter values such as C, γ, and ξ for RBF kernels (Raghavendra and Deka, 2014). It is difficult to optimize these values in SVR. One of the most reliable methods to carry out this work is cross-validation. Also, compared to SVM, there is one more parameter value to be determined in SVR in addition to two parameter values. Various methods have been proposed to speed up model selection. Minimizing bounds of leave-one-out error for SVR is one method to compare with those for classification. Determination of these parameters by altering training of SV regressors and optimizing the parameters by steepest descent for cross-validation data is another method. Based on various performance evaluation of model selection, it was found that k-fold cross-validation, leave-one-out, and span bound worked equally well with different selected values.

4.7.5 Training Methods

When a support vector regressor is expressed by a quadratic optimization problem, the solution is globally optimal. The use of nonlinear kernels needs to solve the dual optimization problem where the number of variables is twice the number of training data. Hence, for a large number of training data, training becomes difficult. To solve this problem, the decomposition technique can be used. The decomposed technique can be classified into the fixed-size chunking. Here, the support vector candidates may exceed the working set size and some of the non-zero elements may move out of the working set. Large-size problems can be trained, but many iterations are required for convergence.

In variable size chunking, usually the number of iterations is smaller compared to fixed-size chunking because of zero elements present in the set. The working set includes variables associated with active constraints where equalities are satisfied. The training methods using variable size chunking developed for pattern classification can be extended to function approximation.

For faster training, Kalman filtering technique is usually used for the sequential training of support vector regressors. Also, training inputs can be positioned on a grid where a matrix associated with the kernel is expressed by a block matrix for faster training.

4.7.6 Variants of SVR

Various variants of SVR are linear programming, ν-SVR, least square, and sparse. For improvement of generalization capability by multiple kernels and outlier detection, enhancement or addition of functions are usually considered. For enhancing approximation ability, a mixture of different kernels is used. For improvement in interpolation and extrapolation abilities, a mix of global kernels, such as polynomials and local kernels, are proposed. Multi-resolution SVR is also proposed based on multi-resolution signal analysis for improvement of performance. Instead of single kernels, multiple kernels with different resolutions are combined. With orthogonal basis functions, components of different resolutions can be computed separately. However, kernels in SVR are not orthogonal, so the support vector training is reformulated to determine the Lagrange multipliers for two types of kernels simultaneously.

4.7.7 Variable Selections

Variable selection is one effective process for reducing computational complexity and improving generalization ability of the regressor. The objective of variable selection is to obtain the smallest set of variables that yields the maximum generalization capability. The wrapper method is time-consuming. And although the filter method reduces computational burden, it is linked with other problems like the risk of a drastic drop in generalization capability in selecting a subset of input variables. To minimize these issues, a combination of both methods for selecting variables during training is proposed called embedded methods.

The variable selection-stopping criteria is one of the important problems in variable selection. Determination of sets of variables with generalization capability is comparable with or better than that of the initial set of variables linked with the selection criterion. The main target is to produce the smallest set of variables with generalization capability compared to the initial set of variables. Hence, before variable selection, a threshold value must be set for

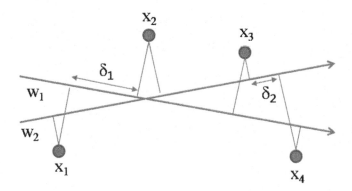

FIGURE 4.6
Linear projection of four data points.

the selection criteria, evaluating approximation error using all the variables. Figure 4.6 shows the linear projection of four data points.

4.7.8 Summary

In this section, optimal hyperplanes, soft margin support vector regression, model selection, training methods, variants of SVR, and variable selection are discussed briefly. The selection of hyperplane related to input–output mappings are explained. Margin for SVR, data set characteristics, and size are discussed in model selection. Various training methods and variants of SVR with selection of variables are explained. Finally, the applicability of the wrapper and filter methods was discussed.

References

Ari, S., Hembram, K., and Saha, G. 2010. Detection of cardiac abnormality from PCG signal using LMS based least square SVM classifier, *Expert Systems with Applications*, 37(12), 8019–8026.

Brooks, J.P. 2011. Support vector machines with the ramp loss and the hard margin loss, *Operations Research*, 59(2), 467–479.

Brunner, C., Fischer, A., Luig, K., and Thies, T. 2012. Pairwise support vector machines and their application to large scale problems, *Journal of Machine Learning Research*, 13(Aug), 2279–2292.

Cortes, C. and Vapnik, V. 1995. Support-vector networks, *Machine Learning*, 20(3), 273–297.

Raghavendra, N.S., Deka, P.C. 2014. Support vector machine applications in the field of hydrology: a review, *Applied Soft Computing* 19, 372–386. doi:10.1016/j. asoc.2014.02.002.

Schölkopf, B. 2001. The kernel trick for distances, In *Advances in Neural Information Processing Systems*, Eds. T.K. Leen, T.G. Dietterich and V. Tresp, MIT Press, 14th Annual Neural Information Processing Systems Conference, Denver, CO pp. 301–307.

Vapnik, V.N. 1995. *The Nature of Statistical Learning Theory*, Springer Verlag, New York.

Wu, Q., and Zhou, D.X. 2005. SVM soft margin classifiers: linear programming versus quadratic programming, *Neural Computation*, 17(5), 1160–1187.

5

Genetic Algorithm (GA)

Genetic Algorithms (GAs) were introduced and developed by John Holland in his book *Adaptation in Natural and Artificial Systems* in 1975 (MIT Press). Holland proposed GA as a heuristic method based on 'survival of the fittest.' GA was discovered as a useful tool for search and optimization problems.

5.1 Introduction

In nature, individuals in a population compete with each other for resources like food, shelter, and so on. Those same individuals also compete to attract mates for reproduction. Due to natural selection, poorly performing individuals have a lesser chance of survival, and the most adapted or 'fit' individuals thrive and produce a relatively large number of offspring. It can also be noted that during reproduction, a recombination of the good characteristics of each ancestor can produce 'fittest' offspring whose fitness is greater than that of a parent. After a few generations, species evolve spontaneously to become more and more adapted to their environment.

Holland described how to apply the principles of natural selection to optimization problems and built the first genetic algorithms (GAs). Holland's theory has been further developed, and now GAs stand up as powerful tools for solving search and optimization problems. Genetic algorithms are based on the principle of genetics and evolution.

The power of mathematics lies in the technology transfer: there exist certain models and methods, which describe many different phenomena and solve a wide variety of problems. GAs are an example of mathematical technology transfer: by simulating evolution, one can solve optimization problems from a variety of sources. Today, GAs are used to resolve complicated optimization problems, like, timetabling, workshop scheduling, and game playing.

5.1.1 Basic Operators and Terminologies in GA

In the following sections we will discuss the basic terminologies and operators used in genetic algorithms to achieve a good enough solution for possible terminating conditions.

Key Elements

Individuals: The two distinct elements in the GA are individuals and populations. An individual is a single solution while the population is the set of individuals currently involved in the search process. Each individual in the population is called a string or *chromosome,* an analogy to chromosomes in natural systems. The chromosome, which is the raw 'genetic' information that the GA deals with. Figure 5.1 shows a chromosome.

Genes: Genes are the basic instructions for building a generic algorithm. A chromosome is a sequence of genes. Genes may describe a possible solution to a problem, without actually being the solution. A gene is a bit of string of an arbitrary length and the location of the gene in a chromosome is called the *locus.* Figure 5.2 shows a gene.

Fitness: The *fitness* function is the evaluation function that is used to determine the fitness of each chromosome. The fitness function is problem specific and user-defined. For calculating fitness, the chromosome has to be first decoded and the objective function has to be evaluated. The fitness not only indicates how good the solution is, but also corresponds to how close the chromosome is to the optimal one.

Population: The population is the set of individuals currently involved in the search. The two important aspects of the population used in genetic algorithms are the initial population generation and the population size. Figure 5.3 shows the population representation.

1	0	1	0	1	0	1	1	1	0	1	1	0

FIGURE 5.1
Representation of a chromosome.

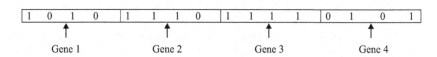

FIGURE 5.2
Representation of gene.

	Chromosome 1	1	1	1	0	0	0	1	0
Population	Chromosome 2	0	1	1	1	1	0	1	1
	Chromosome 3	1	0	1	0	1	0	1	0
	Chromosome 4	1	1	0	0	1	1	0	0

FIGURE 5.3
Population representation.

For each and every problem, the population size will depend on the complexity of the problem. It is often a random initialization of population is carried. Ideally, the first population should have a gene pool as large as possible in order to be able to explore the whole search space.

Encoding: Encoding is a process of representing individual genes. The process can be performed using bits, numbers, trees, arrays, lists, or any other objects. The encoding depends mainly on solving the problem.

Binary Encoding: The most common way of encoding is a binary string, which is shown in Figure 5.4. Each chromosome encodes a binary bit string. Each bit in the string can represent some characteristics of the solution. Every bit string, therefore, is a solution but not necessarily the best solution. Figure 5.4 shows the binary encoding.

Octal Encoding: This encoding uses strings made up of octal numbers (0–7) which are shown in Figure 5.5.

Hexadecimal Encoding: This encoding uses strings made up of hexadecimal numbers (0–9, A–F) as illustrated in Figure 5.6.

Permutation Encoding (Real Number Coding): In permutation encoding, every chromosome is a string of integer/real values, which represents a number in a sequence. Figure 5.7 shows the permutation encoding.

Chromosome 1	1	1	0	1	0	0	0	1	1	0	1	0
Chromosome 2	0	1	1	1	1	1	1	1	1	1	0	0

FIGURE 5.4
Binary encoding.

Chromosome 1	0	3	4	6	7	2	1	6
Chromosome 2	1	5	7	2	3	3	1	4

FIGURE 5.5
Octal encoding.

Chromosome 1	9	C	E	7
Chromosome 2	3	D	B	A

FIGURE 5.6
Hexadecimal encoding.

Chromosome A	1	5	3	2	6	4	7	9	8
Chromosome B	8	5	6	7	2	3	1	4	9

FIGURE 5.7
Permutation (real number) encoding.

Permutation encoding is only useful for ordering problems. Even then, due to crossover and mutation, corrections must be made to ensure the chromosome consistency (i.e., it must have a real sequence in it).

Value Encoding: Every chromosome is a string of values and the values can be anything connected to the problem. This encoding produces the best results for some special problems. In value encoding, every chromosome is a string of some values. Values can be anything connected to problem, form numbers, real numbers or chars, to some complicated objects. On the other hand, for this encoding it is often necessary to develop some new crossover and mutation that is specific to the problem. Value encoding is shown in Figure 5.8.

Breeding (Crossover)

The breeding process is the heart of the genetic algorithm. It is in this process that the search process creates new and hopefully fitter individuals. The breeding cycle consists of three steps:

1. Selecting parents.
2. Crossing the parents to create new individuals (offspring or children).
3. Replacing old individuals in the population with the new ones.

Selection

Selection is the process of choosing two parents from the population for crossing. After deciding on an encoding, the next step is to decide how to perform selection; i.e., how to choose individuals in the population that will create offspring for the next generation and how many offspring each will create. The purpose of selection is to emphasize fitter individuals in the population in the hope that their offspring will have higher fitness. There are different techniques that can be used in GA selection. Some of these techniques are introduced here.

Roulette Wheel Selection: Roulette selection is one of the traditional GA selection techniques. The principle of roulette selection is a linear search through a roulette wheel with the slots in the wheel weighted

Chromosome A	1.2324 5.3243 0.4556 2.3293 2.4545
Chromosome B	A B D J E I F J D H D I E R J F D L D F L F E G T
Chromosome C	(back), (back), (right), (forward), (left)

FIGURE 5.8
Value encoding.

in proportion to the individual's fitness values. The roulette wheel selection process can be explained as follows:

1. The expected value of an individual is that fitness divided by the actual fitness of the population.
2. Each individual is assigned a slice of the roulette wheel, the size of the slice is proportional to the individual's fitness.
3. The wheel is spun N times, where N is the number of individuals in the population.
4. On each spin, the individual under the wheel's marker is selected to be in the pool of parents for the next generation.

Roulette wheel selection is easier to implement, but it is noisy. The rate of evolution depends on the variance of fitness in the population.

Random Selection: This technique randomly selects a parent from the population. Random selection is a little more disruptive, on average, than roulette wheel selection.

Rank Selection: The roulette wheel will have a problem when the fitness values differ very much. If the best chromosome fitness is 90%, its circumference occupies 90% of the roulette wheel, and then other chromosomes have too few chances to be selected. Rank selection ranks the population, and every chromosome receives a fitness value from the ranking. The worst has fitness 1 and the best has fitness N. It results in slow convergence but prevents too quick convergence.

Crossover (Recombination)

Crossover is the process of taking two parent solutions and producing from them a child (offspring). The crossover operator is applied to the mating pool with the hope that it creates better offspring. Crossover is a recombination operator that proceeds in three steps:

1. The reproduction operator selects at random a pair of two individual strings for the mating.
2. A cross site is selected at random along the string length.
3. Finally, the position values are swapped between the two strings following the cross site.

That is, the simplest way to do this is to choose randomly some crossover point and copy everything before this point from the first parent and then copy everything after the crossover point from the other parent.

The various crossover techniques are discussed as follows:

Single-Point Crossover: The traditional genetic algorithm uses a single-point crossover, where the two mating chromosomes are cut once at corresponding points, and the sections after the cuts are exchanged. Here, a cross site or crossover point is selected randomly along the length of the mated strings and bits next to the cross-sites are exchanged. Figure 5.9 illustrates single-point crossover and it can be observed that the bits next to the crossover point are exchanged to produce children. The crossover point can be chosen randomly. Figure 5.9 shows the single-point crossover.

Two-Point Crossover: In two-point crossover, two crossover points are chosen and the contents between these points are exchanged between two mated parents. In Figure 5.10 the dotted lines indicate the crossover points. Thus the contents between these points are exchanged between the parents to produce new children for mating in the next generation. Two-point crossover is shown in Figure 5.10.

Many different crossover algorithms have been devised, often involving more than one cut point. It should be noted that adding further crossover points reduces the performance of the GA.

Uniform Crossover: In uniform crossover each gene in the offspring is created by copying the corresponding gene from one or the other parent chosen according to a randomly generated binary crossover mask of the same length as the chromosomes. Where there is a *1* in the crossover mask, the gene is copied from the first

| Parent 1 | 1 | 0 | 1 | 1 | 0 | 0 | 1 | 0 |
| Parent 2 | 1 | 0 | 1 | 0 | 1 | 1 | 1 | 1 |

| Child 1 | 1 | 0 | 1 | 1 | 0 | 1 | 1 | 1 |
| Child 2 | 1 | 0 | 1 | 0 | 1 | 0 | 1 | 0 |

FIGURE 5.9
Single-point crossover.

| Parent 1 | 1 | 1 | 0 | 1 | 1 | 0 | 1 | 0 |
| Parent 2 | 0 | 1 | 1 | 0 | 1 | 1 | 0 | 0 |

| Child 1 | 1 | 1 | 0 | 0 | 1 | 1 | 1 | 0 |
| Child 2 | 0 | 1 | 1 | 1 | 1 | 0 | 0 | 0 |

FIGURE 5.10
Two point crossover.

parent, and where there is a *0* in the mask, the gene is copied from the second parent.

A new crossover mask is randomly generated for each pair of parents. Offsprings, therefore, contain a mixture of genes from each parent. The number of effective crossing points is not fixed, but will average L/2 (where L is the chromosome length).

In Figure 5.11, it is shown how new children are produced using the uniform crossover approach. While producing child 1 when there is a *1* in the mask, the gene is copied from parent 1 or else from parent 2. On producing child 2, when there is a *1* in the mask, the gene is copied from parent 2; when there is a *0* in the mask, the gene is copied from parent 1.

Crossover Probability: The basic parameter in the crossover technique is the crossover probability (Pc). Crossover probability is a parameter to describe how often crossover will be performed. Without crossover, offspring are exact copies of parents. If there is crossover, offspring are made from parts of both parents' chromosomes. If crossover probability is 100%, then all offspring are made by crossover. If it is 0%, a whole new generation is made from exact copies of chromosomes from the old population. Crossover is used in the hope that new chromosomes will contain good parts of old chromosomes and, therefore, the new chromosomes will be fitter. However, it is good to allow some part of the old population to survive to the next generation.

Mutation: After crossover, the strings are subjected to mutation. The mutation prevents the algorithm from being trapped in a local minimum. It introduces new genetic structures in the population by randomly modifying some of its building blocks. There are many different forms of mutation for the different kinds of representation.

For *binary* representation, a simple mutation can consist in inverting the value of each gene with a small probability. The probability is usually taken about 1/L, where L is the length of the chromosome.

Flipping: Flipping of a bit involves changing 0 to 1 and 1 to 0 based on a mutation chromosome generated.

Parent 1	1	0	1	1	0	0	1	1
Parent 2	0	0	0	1	1	0	1	0
Mask	1	1	0	1	0	1	1	0
Child 1	1	0	0	1	1	0	1	0
Child 2	0	0	1	1	1	0	1	1

FIGURE 5.11
Uniform crossover.

Figure 5.12 explains the mutation-flipping concept. A parent is considered and a mutation chromosome is randomly generated. For a 1 in mutation chromosome, the corresponding bit in parent chromosome is flipped (0 to 1 and 1 to 0) and the child chromosome is produced. In the previous case, 1 occurs at three places of mutation chromosome; the corresponding bits in the parent chromosome are flipped, and the child is generated.

Interchanging: Two random positions of the string are chosen and the bits corresponding to those positions are interchanged. This is shown in Figure 5.13.

Reversing: A random position is chosen, the bits next to that position are reversed, and the child chromosome is produced. This is shown in Figure 5.14.

Mutation probability: The important parameter in the mutation technique is the mutation probability (Pm). The mutation probability decides how often parts of a chromosome will be mutated. If there is no mutation, offspring are generated immediately after crossover (or directly copied) without any change. If mutation is performed, one or more parts of a chromosome are changed. If mutation probability is 100%, the whole chromosome is changed, if it is 0%, nothing is changed. Mutation should not occur very often, because then GA will in fact change to a random search.

Replacement: Replacement is the last stage of any breeding cycle. Two parents are drawn from a fixed size population, they breed two children, but not all four can return to the population, so two must be replaced; i.e., once offspring are produced, a method must determine

Parent	1	0	1	1	0	1	0	1
Mutation Chromosome	1	0	0	0	1	0	0	1
Child	0	0	1	1	1	1	0	0

FIGURE 5.12
Mutation flipping.

Parent	1	0	1	1	0	1	0	1
Child	1	1	1	1	0	0	0	1

FIGURE 5.13
Interchanging.

Parent	1	0	1	1	0	1	0	1
Child	1	0	1	1	0	1	1	0

FIGURE 5.14
Reversing.

which of the current members of the population, if any, should be replaced by the new solutions.

Random Replacement: The children replace two randomly chosen individuals in the population. The parents are also candidates for selection.

Weak Parent Replacement: In weak parent replacement, a weaker parent is replaced by a strong child. With four individuals, only the fittest two—parent or child—return to population. This process improves the overall fitness of the population.

Both Parents: Both parent replacement is simple. The child replaces the parent. In this case, each individual only gets to breed once. As a result, the population and genetic material moves around but leads to a problem when combined with a selection technique that strongly favors fit parents: the fit breed, and then they are disposed of.

Elitism: The first best chromosome or the few best chromosomes are copied to the new population. The rest is done in a classical way. Such individuals can be lost if they are not selected to reproduce or if crossover or mutation destroys them. This significantly improves the GA's performance.

5.1.2 Traditional Algorithm and GA

The principle of GAs is simple: imitate genetics and natural selection by a computer program. The parameters of the problem are coded most naturally as a DNA-like linear data structure, a vector, or a string. Sometimes, when the problem is naturally two- or three-dimensional, other corresponding array structures are used.

A set, called population, which is a dependent parameter value vector, is processed by GA. To start there is usually a totally random population, the values of different parameters generated by a random number generator. Typical population size is from a few dozen to thousands. To do optimization, we need a cost function (or fitness function, as it is usually called) when genetic algorithms are used. By using a fitness function, we can select the best solution (candidates) from the population and delete the inferior specimens.

The nice thing when comparing GAs to other optimization methods is that the fitness function can be nearly anything that can be evaluated by a computer or even something that cannot! In the latter case, it might be a human judgment that cannot be stated as a crisp program, as in the case of an eyewitness, where a human being selects among the alternatives generated by GA.

So, there are not any definite mathematical restrictions on the properties of the fitness function. It may be discrete, multimodal, etc.

The main criteria used to classify optimization algorithms are as follows: continuous/discrete, constrained/unconstrained, and sequential/parallel.

There is a clear difference between discrete and continuous problems. Therefore, it is instructive to notice that continuous methods are sometimes used to solve inherently discrete problems and vice versa. Parallel algorithms are usually used to speed up processing. There are, however, some cases in which it is more efficient to run several processors in parallel rather than sequentially. These cases include, among others, those in which there is high probability of each individual search run getting stuck in a local extreme.

Irrespective of the above classification, optimization methods can be further classified into deterministic and non-deterministic methods. In addition, optimization algorithms can be classified as local or global. In terms of energy and entropy, local search corresponds to entropy while global optimization depends essentially on fitness, i.e. energy landscape.

The genetic algorithm differs from conventional optimization techniques in the following ways:

- GAs operate with coded versions of the problem parameters rather than parameters themselves. That is, GA works with the coding of the solution set and not with the solution itself.

- Almost all conventional optimization techniques search from a single point but GAs always operate on a whole population of points (strings); i.e., GA uses a population of solutions rather than a single solution from searching. This plays a major role in the robustness of genetic algorithms. It improves the chance of reaching the global optimum and also helps in avoiding a local stationary point.

- GA uses the fitness function for evaluation rather than derivatives. As a result, they can be applied to any kind of continuous or discrete optimization problem. The key point to be performed here is to identify and specify a meaningful decoding function.

- GAs use probabilistic transition operates, while conventional methods for continuous optimization apply deterministic transition operates; i.e., GAs do not use deterministic rules.

5.1.3 General GA

An algorithm is a series of steps for solving a problem. A genetic algorithm is a problem-solving method that uses genetics as its model of problem solving. It's a search technique to find approximate solutions to optimization and search problems.

Basically, an optimization problem looks really simple. One knows the form of all possible solutions corresponding to a specific question. The set of all the solutions that meet this form constitute the search space. The problem consists of finding out the solution that fits the best (i.e., the one with the most payoffs) from all the possible solutions. If it's possible to quickly enumerate all the solutions, the problem does not raise much difficulty. But, when the search space becomes large, the enumeration is soon no longer feasible simply because it would take far too much time. That's when a specific technique is needed to find the optimal solution. Genetic Algorithms provides one of these methods. Practically, they all work in a similar way, adapting simple genetics to algorithmic mechanisms.

GA handles a population of possible solutions. Each solution is represented through a chromosome, which is just an abstract representation. Coding all the possible solutions into a chromosome is the first part, but certainly not the most straightforward one of a genetic algorithm. A set of reproduction operators has to be determined, too. Reproduction operators are applied directly to the chromosomes and are used to perform mutations and recombinations over solutions of the problem. Appropriate representation and reproduction operators are really something determinant, as the behavior of the GA is extremely dependent on it. Frequently, it can be extremely difficult to find a representation, which respects the structure of the search space and reproduction operators, which are coherent and relevant according to the properties of the problems.

Selection is supposed to be able to compare each individual in the population. Selection is done by using a fitness function. Each chromosome has an associated value corresponding to the fitness of the solution it represents. The fitness should correspond to an evaluation of how good the candidate solution is. The optimal solution is the one that maximizes the fitness function.

Genetic algorithms deal with the problems that maximize the fitness function. But if the problem consists in minimizing a cost function, the adaptation is quite easy. Either the cost function can be transformed into a fitness function, for example by inverting it; or the selection can be adapted in such way that they consider individuals with low evaluation functions to be superior.

Once the reproduction and the fitness function have been properly defined, a genetic algorithm is evolved according to the same basic structure. It starts by generating an initial population of chromosomes. This first population must offer a wide diversity of genetic material. The gene pool should be as large as possible so that any solution of the search space can be engendered.

Generally, the initial population is generated randomly. Then, the genetic algorithm loops over an iteration process to make the population evolve. Each iteration consists of the following steps:

1. SELECTION: The first step consists in selecting individuals for reproduction. This selection is done randomly with a probability depending on the relative fitness of the individuals so that best ones are more often chosen for reproduction than the poorer ones.

2. REPRODUCTION: In the second step, offspring are bred by the selected individuals. For generating new chromosomes, the algorithm can use both recombination and mutation.

3. EVALUATION: Then the fitness of the new chromosomes is evaluated.

4. REPLACEMENT: During the last step, individuals from the old population are killed and replaced by the new ones.

The algorithms stop when the population converges toward the optimal solution.

The basic genetic algorithm is as follows:

- [Start] Genetic random population of n chromosomes (suitable solutions for the problem)
- [Fitness] Evaluate the fitness f(x) of each chromosome x in the population
- [New population] Create a new population by repeating the following steps until the new population is complete
  - [Selection] Select two parent chromosomes from a population according to their fitness (the better fitness, the greater the chance of being selected).
  - [Crossover] With crossover probability, cross over the parents to form new offspring (children). If no crossover was performed, the offspring will be an exact copy of the parents.
  - [Mutation] With mutation probability, mutate new offspring at each locus (position in chromosome).
  - [Accepting] Place new offspring in the new population.
  - [Replace] Use a newly generated population for a further sum of the algorithm.
- [Test] If the end condition is satisfied, stop and return the best solution in current population.
- [Loop] Go to Step 2 for fitness evaluation.

The genetic algorithm process is discussed through the GA cycle in Figure 5.15.

Reproduction is the process by which the genetic material in two or more parents is combined to obtain one or more offspring. In the fitness evaluation step, the individual's quality is assessed. Mutation is performed to one individual to produce a new version of it where some of the original genetic material has been randomly changed. The selection process helps to decide which individuals are to be used for reproduction and mutation in order to produce new search points.

The flowchart showing the process of GA is shown in Figure 5.16.

Before implementing GAs, it is important to understand a few guidelines for designing a general search algorithm (i.e., a global optimization

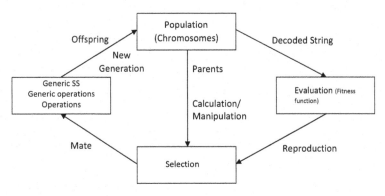

FIGURE 5.15
Genetic algorithm cycle.

algorithm) based on the properties of the fitness landscape. The most common optimization methods are:

1. **Determinism:** A purely deterministic search may have an extremely high variance in solution quality, because it may soon get stuck in worst-case situations from which it is impossible to escape due to its determinism. This can be avoided, but it is a well-known fact that the observation of the worst-case situation is not guaranteed to be possible in general.

2. **Non-determinism:** A stochastic search method usually does not suffer from the above potential worst-case 'wolf trap' phenomenon. It is, therefore, likely that a search method should be stochastic, but it may well contain a substantial portion of determinism. In principle, it is enough to have as much non-determinism as to be able to avoid the worst-case wolf traps.

3. **Local Determinism:** A purely stochastic method is usually quite slow. It is therefore reasonable to do as much as possible using efficient deterministic predictions of the most promising directions of (local) proceedings. This is called local hill climbing or greedy search according to the obvious strategies.

Based on the foregoing discussion, the important criteria for GA approach can be formulated as given below:

- Completeness: Any solution should have its encoding.
- Non-redundancy: Codes and solutions should correspond one to one another.
- Soundness: Any code (produced by genetic operators) should have its corresponding solution.
- Characteristic perseverance: Offspring should inherit useful characteristics from parents.

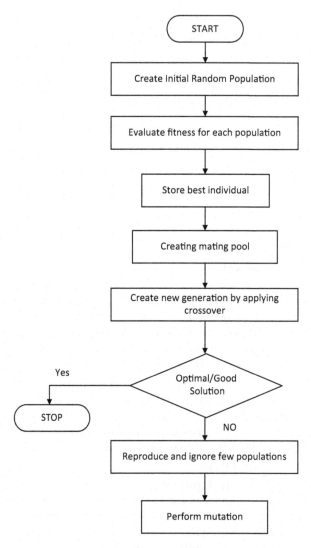

FIGURE 5.16
Flowchart of genetic algorithm.

In short, the basic four steps used in simple genetic algorithms to solve problems are:

- The representation of the problem
- The fitness calculation
- Various variables and parameters involved in controlling the algorithm
- The representation of result and the way of terminating the algorithm

5.1.4 The Schema Theorem

A schema is a similarity template describing a subset of a string displaying similarities at certain string positions. It is formed by the ternary alphabet {0.1,*}, where * is simply a notation symbol that allows the description of all possible similarities among strings of a particular length and alphabet. In general, there are 2^1 different strings or chromosomes of length 1, but schemata display an order of 3^1. A particular string of length 1 inside a population of 'n' individuals into one of the 2^1 schemata can be obtained from this string. Thus, in the entire population the number of schemata present in each generation is somewhere between 2^1 and $n.2^1$, depending upon the population diversity. Holland estimated that in a population of 'n' chromosomes, the GAs process $O(n^3)$ schemata into each generation. This is called implicit parallel process.

A schema represents an affined variety of the search space: for example, the schema 01**11*0 is a sub-space of the space of codes of 8-bits length (* can be 0 or 1).

The GA modeled in schema theory is a canonical GA, which acts on binary strings and for which the creation of a new generation is based on three operators:

- A proportionate selection, where the fitness function steps in: the probability that a solution of the current population is selected and is proportional to its fitness.
- The genetic operators: single-point crossover and third operator is a bit-flip mutation, randomly applied with probabilities Pc and Pm.

Schemata represent global information about the fitness function. A GA works on a population of N codes, and implicitly uses information on a certain number of schemata. The basic 'schema theorem' presented below is based on the observation that the evaluation of a single code makes it possible to deduce some knowledge about the schemata to which that code belongs.

Theorem: Schema Theorem (Holland)

The Schema Theorem is called the 'fundamental theorem of genetic algorithms.'

For a given schema H, let:

- m (H, t) be the relative frequency of the schema H in the population of the tth generation.
- f(H) be the mean fitness of the elements of H.
- O(H) be the number of fixed bits in the schema H, called the order of the schema.
- δ(H) be the distance between the first and the last fixed bit of the schema, called the definition length of the schema.

- f is the mean fitness of the current population.
- P_c is the crossover probability.
- P_m is the mutation probability.

Then,

$$E[m(H, t+1)] \geq m(H, t)\frac{f(H)}{\bar{f}}\left[1 - P_c\frac{\delta(H)}{1-1} - O(H)P_m\right] \qquad (5.1)$$

Based on a qualitative view, the above formula means that the 'good' schemata, having a short definition length and a low order, tend to grow very rapidly in the population. These particular schemata are called building blocks.

The application of the Schema Theorem is as follows:

- It provides some tools to check whether a given representation is well-suited to a GA.
- The analysis of nature of the 'good' schemata gives few ideas on the efficiency of genetic algorithm.

5.1.5 Optimal Allocation of Trails

The Schema Theorem has provided the insight that building blocks receive exponentially increasing trials in future generations. This leads to an important and well-analyzed problem from statistical decision theory—the two-armed bandit problem and its generalization, the k-armed bandit problem.

Consider a gambling machine with two slots for coins and two arms. The gambler can deposit the coin either into the left or the right slot. After pulling the corresponding arm, either a reward is paid, or the coin is lost. For mathematical simplicity, working only with outcomes, i.e., the difference between the reward (which can be zero) and the value of the coin. Let us assume that the left arm produces an outcome with mean value μ_1 and a variance $\sigma21$ while the right arm produces an outcome with mean value μ_2 and variance $\sigma22$. Without loss of generality, although the gambler does not know this, assume that $\mu_1 \geq \mu_2$.

The question arises which arm should be played. Since it is not known beforehand which arm is associated with the higher outcome, we are faced with an interesting dilemma. Not only must the gambler make a sequence of decisions (like which arm to play), the gambler must also, at the same time, collect information about which is the better arm. This trade-off between exploration of knowledge and its exploitation is the key issue in this problem and, as turns out later, in genetic algorithms, too.

A simple approach to this problem is to separate exploration from exploitation. More specifically, it is possible to perform a single experiment at the beginning and thereafter make an irreversible decision that depends on the results of the experiment.

Suppose we have N coins. If we first allocate an equal number n (where $2n \leq N$) of trials to both arms, we could allocate the remaining $N - 2n$ trials to the observed better arm. Assuming we know all involved parameters, the expected loss is given as

$$L(N,n) = (\mu_1 - \mu_2) \cdot ((N-n)q(n) + n(1 - q(n))) \tag{5.2}$$

where $q(n)$ is the probability that the worst arm is the observed best arm after the $2n$ experimental trials. The underlying idea is obvious: in case we observe that the worse arm is the best, which happens with probability $q(n)$, the total number of trials allocated to the right arm is $N - n$. The loss is, therefore, $(\mu_1 - \mu_2) \cdot (N - n)$. In the reverse case that we actually observe that the best arm is the best, which happens with probability $1 - q(n)$, the loss is only what we get less because we played the worse arm n times, i.e., $(\mu_1 - \mu_2) \cdot n$. Taking the central limit theorem into account, we can approximate $q(n)$ with the tail of a normal distribution:

$$q(n) \approx \frac{1}{\sqrt{2\pi}} \cdot \frac{c^{-e^2/2}}{c}, \quad \text{where} \quad c = \frac{\mu_1 - \mu_2}{\sqrt{\sigma_1^2 + \sigma_2^2}} \cdot \sqrt{n} \tag{5.3}$$

Now we have to specify a reasonable experiment size n. Obviously, if we choose $n = 1$, the obtained information is potentially unreliable. If we, however, choose $n = N/2$, there are no trials left to make use of the information gained through the experimental phase. What we see is again the trade-off between exploitation with almost no exploration ($n = 1$) and exploration without exploitation ($n = N/2$). It does not take a Nobel prize winner to see that the optimal way is somewhere in the middle.

A genetic algorithm, although the direct connection is not yet fully clear, actually comes close to this ideal, giving at least an exponentially increasing number of trials to the observed best building blocks. However, one may still wonder how the two-armed bandit problem and GAs are related.

Let us consider an arbitrary string position. Then there are two schemata of order one, which have their only specification in this position. According to the Schema Theorem, the GA implicitly decides between these two schemata, where only incomplete data are available (observed average fitness values). In this sense, a GA solves a lot of two-armed problems in parallel.

The Schema Theorem, however, is not restricted to schemata with an order of 2. Looking at competing schemata (different schemata which are specified in the same positions), we observe that a GA is solving an enormous number of k-armed bandit problems in parallel. The k-armed bandit problem, although much more complicated, is solved in an analogous way—the observed better alternatives should receive an exponentially increasing number of trials.

5.1.6 Summary

In this section, various terminologies and basic operators are briefly discussed. Traditional algorithm and contrast with GA is discussed. The need for schemata approach is also discussed. The optimal allocation of trials is introduced.

5.2 Classification of GA

5.2.1 Introduction

Genetic algorithms are search algorithms based on the mechanics of natural selection and natural genetics. Algorithms are nothing but step-by-step procedures to find solutions to problems.

Genetic algorithms also give the step-by-step procedure to solve the problem, but they are based on the genetic models. Genetic algorithms are theoretically and empirically proven to provide robust search in complex phases with the previously mentioned features. Genetic algorithms are capable of giving rise to efficient and effective searches in the problem domain and they are now finding more widespread application in business, science, and engineering. These algorithms are computationally less complex but more powerful in their search for improvement. These features have enabled researchers to form different approaches of genetic algorithms.

5.2.2 Adaptive GA

Adaptive genetic algorithms (AGA) are GAs whose parameters, such as the population size, the crossing over probability, or the mutation probability are varied while the GA is running. A simple variant could be the following: the mutation rate is changed according to changes in the population; the longer the population does not improve, the higher the mutation rate is chosen. Vice versa, it is decreased again as soon as an improvement of the population occurs. The overall adaptive genetic algorithm procedure follows these steps:

Step 1: Initial population.

We use the population obtained by random number generation.

Step 2: Genetic operators.

Selection: elitist strategy in an enlarged sampling space

Crossover: order-based crossover operator for activity priority

Mutation, local search-based mutation operator for activity mode

Step 3: Apply the local search using iterative hill-climbing method in GA loop.

Step 4: Apply the heuristic for adaptively regulating GA parameters (i.e., the rates of crossover and the mutation operators).

Step 5: Stop condition.

If a predefined maximum generation number is reached or an optimal solution is located during genetic search process, then stop; otherwise, go to Step 2.

For a multimode function, if it is needed to keep the global search ability it must have balanced searchability. Crossover probability (Pc) and mutation probability (Pm) are the main factors in affecting balanced search ability (global search ability and local searchability). While we strengthen one ability by increasing or decreasing Pc and Pm, we may weaken other abilities. Both Pc and Pm in the simple genetic algorithm (SGA) are invariant, so, for the complex optimal problem, the GA's efficiency is not high. In addition, immature convergence may be caused.

Therefore, the goals with adaptive probabilities of crossover and mutation are to maintain the genetic diversity in the population and prevent the genetic algorithms from converging prematurely to local minima. As a result, the adaptive genetic algorithm was developed, and its basic idea is to adjust Pc and Pm according to the individual fitness. This algorithm can better solve the problem of adjusting Pc and Pm dynamically and also fits to all kinds of optimal problems.

5.2.3 Hybrid GA

A hybrid genetic algorithm has been designed by combining a variant of an already existing crossover operator with these heuristics. One of the heuristics is for generating initial population; two others are applied to the offspring either obtained by crossover or by shuffling. The last two heuristics applied to offspring are greedy in nature; hence, to prevent getting stuck on local optimum, one has to include the proper amount of randomness by using the shuffling operator.

The hybrid genetic algorithm in this section is designed to use heuristics for initialization of population and improvement of offspring produced by crossover for a traveling salesman problem (TSP). The initialization heuristics algorithm is used to initialize a part of the population; the remaining part of the population will be initialized randomly. The offspring is obtained by crossover between two parents selected randomly. The tour improvement heuristics, *RemoveSharp* and *LocalOpt*, are used to bring the offspring to a local minimum. If the cost of the tour of the offspring thus obtained is less than the cost of the tour of any one of the parents, then the parent with higher cost is removed from the population and the offspring is added to the population. If the cost of the tour of the offspring is greater than that of both of its parent, then it is discarded. For shuffling, a random number is generated within one and if it is less than the specified probability of the shuffling operator, a tour is randomly selected and is removed from the population. Its sequence is randomized and then added to the population.

5.2.4 Parallel GA

Parallel GAs (PGAs) are relatively easy to implement and assure substantial gains in performance. The major aspect of GAs is its ability to be parallelized. As natural evolution deals with an entire population, and not only with particular individuals, it is a significantly high parallel process. Except in the selection phase, during which competition may take place between individuals, the only interactions between members of the population occur during reproduction phase. Usually not more than two individuals are necessary to engender a new generation child. For any other operations of the evolution, evaluation of each member of the population can be done separately. Hence, almost all the operations in a GA are implicitly parallel.

PGAs normally work in distributing the task of a basic GA on different processors. When these tasks are implicitly parallel, less time will be spent on communication and the algorithm is expected to run much faster or find more accurate results.

The efficiency of GA in finding optimal solutions is mostly determined by the population size. As size of population increases, the genetic diversity also increases. Hence, the algorithm takes more time to converge as a large amount of memory needs to be stored.

In today's world, new parallel computers not only provide more space but also allow the use of several processors to evaluate and produce more solutions in less time. With the emergence of new high-performance computing, researchers are heavily focused on improving the performance of GAs.

5.2.5 Messy GA

Messy GA (MGA) is introduced to overcome the difficulties of classical GA, where genes are encoded in a fixed order. Usually GA is likely to converge well if the optimization task can be divided into several short building blocks. Again, one-point crossover tends to disadvantage long schemata over short ones. These limitations are overcome by using a variable length, position-independent coding. The MGA concept is to append an index to each gene which allows for the identification of its position. It can identify the genes uniquely with the help of the index as genes may be swapped arbitrarily without changing the meaning of the string. The MGA normally works with the same mutation operator as a simple GA; the crossover operator is replaced by a more general cut and splice operator, which also allows parents with different lengths to mate.

5.2.6 Real Coded GA

The real coded GAs (RGAs) are related to the variant of GA for real-valued optimization that is closest to the original GA. In an RGA, an individual is represented as an N-dimensional vector of real numbers. This RGA deals with free N-dimensional real-valued optimization problems. The selection does not involve the particular coding, and no adaptation needs to be made without any restriction. The only adaptations are the genetic operations: crossover and mutation.

5.2.7 Summary

In this section, various classes of GA have been discussed. The class of parallel GA is very complex as its behavior is affected by many parameters. The MGA was introduced to improve efficiency and minimize some difficulties in GA. Real coded GA is getting attention due to the real-valued optimization problem. Also, the hybrid GA and adaptive GA have been discussed with the necessary information.

5.3 Genetic Programming

One of the central challenges of computer science is to get a computer to do what needs to be done without telling it how to do it. Genetic programming (GP) addresses this challenge by providing a method for automatically creating a working computer program from a high-level problem statement of the problem. Genetic programming achieves this goal of *automatic programming* (also called *program synthesis* or *program induction*) by genetically breeding a population of computer programs using the principles of Darwinian natural selection and biologically inspired operations. The operations include reproduction, crossover (sexual recombination), mutation, and architecture-altering operations patterned after gene duplication and gene deletion in nature. For example, an element of a population might correspond to an arbitrary placement of eight queens on a chessboard, and the fitness function might count the number of queens that are not attacked by any other queens. Given an appropriate set of genetic operators by which an initial population of queen placements can spawn new collections of queen placements, a suitably designed system could solve the classic 'eight queens' problem.

GP's uniqueness comes from the fact that it manipulates populations of structured programs—in contrast to much of the work in evolutionary computation in which population elements are represented using flat strings over some alphabet. In this chapter, the basic concepts— working, representations, and applications of genetic programming—have been dealt with in detail.

5.3.1 Introduction

Genetic programming (GP) is a domain independent, problem-solving approach in which computer programs are evolved to find solutions to problems. The solution technique is based on the Darwinian principle of 'survival of the fittest' and is closely related to the field of genetic algorithms (GA).

However, three important differences exist between GAs and GP:

- **Structure:** GP usually evolves tree structures while GAs evolve binary or real number strings.
- **Active vs passive:** Because GP usually evolves computer programs, the solutions can be executed without post-processing (i.e., active structures), whereas GAs typically operate on coded binary strings (i.e., passive structures), which require post-processing.
- **Variable vs fixed length:** In traditional GAs, the length of the binary string is fixed before the solution procedure begins. However, a GP parse tree can vary in length throughout the run. (Although it is recognized that in more advanced GA work, variable length strings are used.)

The ability to search the solution space and locate regions that potentially contain optimal solutions for a given problem is one of the fundamental components of most artificial intelligence (AI) systems. There are three primary types of search: the blind search, hill climbing, and beam search. GP is classified as a beam search because it maintains a population of solutions that is smaller than all of the available solutions. GP is also usually implemented as a weak search algorithm as it contains no problem-specific knowledge, although some research has been directed towards 'strongly typed genetic programming.' However, while GP can find regions containing optimal solutions, an additional local search algorithm is normally required to locate the optima. Memetic algorithms can fulfill this role by combining an evolutionary algorithm with problem-specific search algorithm to locate optimal solutions.

5.3.2 Characteristics of GP

Genetic programming now delivers High-Return Human-Competitive Machine Intelligence.

Based on this sentence, it can be noted that the four main characteristics of genetic programming are:

- human-competitive
- high-return
- routine
- machine intelligence

5.3.2.1 Human-Competitive

In attempting to evaluate an automated problem-solving method, the question arises as to whether there is any real substance to the demonstrative problems that are published in connection with the method. Demonstrative problems in the fields of artificial intelligence and machine learning are often contrived toy problems that circulate exclusively inside academic groups that study a particular methodology.

These problems typically have little relevance to any issues pursued by any scientist or engineer outside the fields of artificial intelligence and machine learning. Indeed, getting machines to produce human-like results is *the* reason for the existence of the fields of artificial intelligence and machine learning.

5.3.2.2 High-Return

What is delivered by the actual automated operation of an artificial method in comparison to the amount of knowledge, information, analysis, and intelligence that is pre-supplied by the human employing the method? The *AI ratio* (the 'artificial-to-intelligence' ratio) of a problem-solving method as the ratio of that which is delivered by the automated operation of the *artificial* method to the amount of *intelligence* that is supplied by the human applying the method to a particular problem.

The AI ratio is especially pertinent to methods for getting computers to automatically solve problems because it measures the value added by the artificial problem-solving method. Manifestly, the aim of the fields of artificial intelligence and machine learning is to generate human-competitive results with a high AI ratio.

The aim of the fields of artificial intelligence and machine learning is to get computers to automatically generate human-competitive results with a high AI ratio—not to have humans generate human-competitive results themselves.

5.3.2.3 Routine

Generality is a precondition that an automated problem-solving method is 'routine.' Once the generality of a method is established, 'routineness' means that relatively little human effort is required to get the method to successfully handle new problems within a particular domain and to successfully handle new problems from a different domain.

A problem-solving method cannot be considered routine if its executional steps must be substantially augmented, deleted, rearranged, reworked, or customized by the human user for each new problem.

5.3.2.4 Machine Intelligence

Since the beginning, the terms *machine intelligence*, *artificial intelligence*, and *machine learning* all referred to the goal of getting 'machines to exhibit behavior, which if done by humans, would be assumed to involve the use of intelligence.' However, in the intervening five decades, the terms *artificial intelligence* and *machine learning* progressively diverged from their original goal-oriented meaning. These terms are now primarily associated with particular methodologies for attempting to achieve the goal of getting computers to automatically solve problems.

Thus, the term *artificial intelligence* is today primarily associated with attempts to get computers to solve problems using methods that rely on knowledge, logic, and various analytical and mathematical methods. The term *machine learning* is today primarily associated with attempts to get computers to solve problems that use a particular small and somewhat arbitrarily chosen set of methodologies (many of which are statistical in nature).

The second approach for achieving machine intelligence was in which previously acquired knowledge is accumulated, stored in libraries, and brought to bear in solving a problem—the approach taken by modern knowledge-based expert systems.

5.3.3 Working of GP

The steps of genetic programming are:

- Preparatory steps
- Executional steps

5.3.3.1 Preparatory Steps of Genetic Programming

The human user communicates the high-level statement of the problem to the genetic programming system by performing certain well-defined preparatory steps.

The five major preparatory steps for the basic version of genetic programming require the human user to specify

(1) the set of terminals (e.g., the independent variables of the problem, zero argument functions, and random constants) for each branch of the to-be-evolved program;

(2) the set of primitive functions for each branch of the to-be-evolved program;

(3) the fitness measure (for explicitly or implicitly measuring the fitness of individuals in the population);

(4) certain parameters for controlling the run; and

(5) the termination criterion and method for designating the result of the run.

The preparatory steps are the problem-specific and domain-specific steps that are performed by the human user prior to launching a run of the problem-solving method. Figure 5.17 shows the five major preparatory steps for the basic version of genetic programming. The preparatory steps (shown at the top of the figure) are the input to the genetic programming system. A computer program (shown at the bottom) is the output of the genetic programming system. The program that is automatically created by genetic programming may solve, or approximately solve, the user's problem.

Genetic programming requires a set of primitive ingredients to get started. The first two preparatory steps specify the primitive ingredients that are to be used to create the to-be-evolved programs. The universe of allowable compositions of these ingredients defines the search space for a run of genetic programming. Figure 5.17 shows the preparatory steps of genetic programming.

The identification of the function set and terminal set for a particular problem (or category of problems) is often a mundane and straightforward process that requires only the minimum knowledge and related information about the problem domain.

For example, if the goal is to get genetic programming to automatically program a robot to mop the entire floor of an obstacle-laden room, the human user must tell genetic programming that the robot is capable of executing functions such as moving, turning, and swishing the mop.

The human user must supply this information prior to a run because the genetic programming system does not have any built-in knowledge telling it that the robot can perform these particular functions. Of course, the necessity of specifying a problem's primitive ingredients is not a unique requirement of genetic programming.

It would be necessary to impart this same basic information to a neural network learning algorithm, a reinforcement-learning algorithm, a decision tree, a classifier system, an automated logic algorithm, or virtually any other automated technique that is likely to be used to solve this problem.

Similarly, if genetic programming is to automatically synthesize an analog electrical circuit, the human user must supply basic information about the

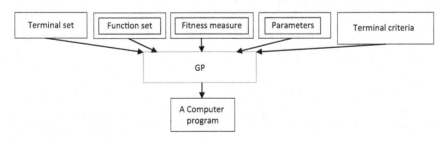

FIGURE 5.17
Preparatory steps of genetic programming.

ingredients that are appropriate for solving a problem in the domain of ana-log circuit synthesis.

In particular, the human user must inform genetic programming that the components of the to-be-created circuit may include transistors, capacitors, and resistors (as opposed to, say, neon bulbs, relays, and doorbells). Although this information may be second nature to anyone working with electrical circuits, genetic programming does not have any built-in knowledge con-cerning the fact that transistors, capacitors, and resistors are the workhorse components for nearly all present-day electrical circuits.

Once the human user has identified the primitive ingredients, the same function set can be used to automatically synthesize amplifiers, computa-tional circuits, active filters, voltage reference circuits, and any other circuit composed of these basic ingredients.

Likewise, genetic programming does not know that the inputs to a con-troller include the reference signal and plant output and that controllers are composed of integrators, differentiators, leads, lags, gains, adders, subtrac-tors, and the like. Thus, if genetic programming is to automatically synthe-size a controller, the human user must give genetic programming this basic information about the field of control.

The third preparatory step concerns the fitness measure for the problem. The fitness measure specifies what needs to be done. The result that is pro-duced by genetic programming specifies 'how to do it.' The fitness measure is the primary mechanism for communicating the high-level statement of the problem's requirements to the genetic programming system. If one views the first two preparatory steps as defining the search space for the problem, one can then view the third preparatory step (the fitness measure) as specify-ing the search's desired direction.

The fitness measure is the means of ascertaining that one candidate indi-vidual is better than another. That is, the fitness measure is used to establish a partial order among candidate individuals.

The partial order is used during the executional steps of genetic program-ming to select individuals to participate in the various genetic operations (i.e., crossover, reproduction, mutation, and the architecture-altering opera-tions). The fitness measure is derived from the high-level statement of the problem. Indeed, for many problems, the fitness measure may be almost identical to the high-level statement of the problem. The fitness measure typically assigns a single numeric value reflecting the extent to which a candidate individual satisfies the problem's high-level requirements. For example:

- If an electrical engineer needs a circuit that amplifies an incoming signal by a factor of 1,000, the fitness measure might assign fitness to a candidate circuit based on how closely the circuit's output comes to a target signal whose amplitude is 1,000 times that of the incoming signal. In comparing two candidate circuits, amplification of 990:1 would be considered better than 980:1.

- If a control engineer wants to design a controller for the cruise control device in a car, the fitness measure might be based on the time required to bring the car's speed up from 55 to 65 miles per hour. When candidate controllers are compared, a rise time of 10.1 seconds would be considered better than 10.2 seconds.

- If a robot is expected to mop a room, the fitness measure might be based on the percentage of the area of the floor that is cleaned within a reasonable amount of time.

- If a classifier is needed for protein sequences (or any other objects), the fitness measure might be based on the correlation between the category to which the classifier assigns each protein sequence and the correct category.

- If a biochemist wants to find a network of chemical reactions or a metabolic pathway that matches observed data, the fitness measure might assign fitness to a candidate network based on how closely the network's output matches the data.

The fitness measure for a real-world problem is typically multi-objective. That is, there may be more than one element that is considered in ascertaining fitness. For example, the engineer may want an amplifier with 1,000:1 gain, but may also want low distortion, low bias, and a low parts count. In practice, the elements of a multi-objective fitness measure usually conflict with one another. Thus, a multi-objective fitness measure must prioritize the different elements so as to reflect the trade-offs that the engineer is willing to accept. For example, the engineer may be willing to tolerate an additional 1% of distortion in exchange for the elimination of one part from the circuit. One approach is to blend the distinct elements of a fitness measure into a single numerical value (often merely by weighting them and adding them together).

The fourth and fifth preparatory steps are administrative. The fourth preparatory step entails specifying the control parameters for the run. The major control parameters are the population size and the number of generations to be run. Some analytic methods are available for suggesting optimal population sizes for runs of the genetic algorithm on particular problems. However, the practical reality is that, generally, it will not use any such analytic method to choose the population size. Instead, it determines the population size in such a way that genetic programming can execute a reasonably large number of generations within the amount of computer time that relate to devote to the problem. As for other control parameters, broadly speaking, it uses the same (undoubtedly non-optimal) set of minor control parameters from problem to problem over a period of years. The fifth preparatory step consists of specifying the termination criterion and the method of designating the result of the run.

5.3.3.2 Executional Steps of Genetic Programming

Genetic programming typically starts with a population of randomly generated computer programs composed of the available programmatic ingredients. Genetic programming iteratively transforms a population of computer programs into a new generation of the population by applying analogs of naturally occurring genetic operations. These operations are applied to individual(s) selected from the population.

The individuals are probabilistically selected to participate in the genetic operations based on their fitness (as measured by the fitness measure provided by the human user in the third preparatory step). The iterative transformation of the population is executed inside the main generational loop of the run of genetic programming.

The executional steps of genetic programming (that is, the flowchart of genetic programming) are as follows:

(1) Randomly create an initial population (generation 0) of individual computer programs composed of the available functions and terminals.

(2) Iteratively perform the following sub-steps (called a *generation*) on the population until the termination criterion is satisfied:

    (a) Execute each program in the population and ascertain its fitness (explicitly or implicitly) using the problem's fitness measure.

    (b) Select one or two individual program(s) from the population with a probability based on fitness (with re-selection allowed) to participate in the genetic operations in (c).

    (c) Create new individual program(s) for the population by applying the following genetic operations with specified probabilities:

        (i) *Reproduction*: Copy the selected individual program to the new population.

        (ii) *Crossover*: Create new offspring program(s) for the new population by recombining randomly chosen parts from two selected programs.

        (iii) *Mutation*: Create one new offspring program for the new population by randomly mutating a randomly chosen part of one selected program.

        (iv) *Architecture-altering operations*: Choose an architecture-altering operation from the available repertoire of such operations and create one new offspring program for the new population by applying the chosen architecture-altering operation to one selected program.

(3) After the termination criterion is satisfied, the single best program in the population produced during the run (the best-so-far individual) is harvested and designated as the result of the run. If the run is successful, the result may be a solution (or approximate solution) to the problem.

5.3.3.3 Fitness Function

The most difficult and most important concept of genetic programming is the fitness function. The fitness function determines how well a program is able to solve the problem. It varies greatly from one type of program to the next. For example, if one were to create a genetic program to set the time of a clock, the fitness function would simply be the amount of time that the clock is wrong. Unfortunately, few problems have such an easy fitness function; most cases require a slight modification of the problem in order to find the fitness.

5.3.3.4 Functions and Terminals

The terminal and function sets are also important components of genetic programming. The terminal and function sets are the alphabet of the programs to be made. The terminal set consists of the variables and constants of the programs. In some cases, the terminal set would contain three commands: forward, right and left.

The function set consists of the functions of the program. In a few cases, the function set would contain: if 'dot,' then do x, else do y. The functions are several mathematical functions, such as addition, subtraction, division, multiplication, and other, more complex functions.

5.3.3.5 Crossover Operation

Two primary operations exist for modifying structures in genetic programming. The most important one is the crossover operation. In the crossover operation, two solutions are sexually combined to form two new solutions or offspring. The parents are chosen from the population by a function of the fitness of the solutions. Three methods exist for selecting the solutions for the crossover operation: probability, tournament, and rank. The first method uses probability based on the fitness of the solution. If $f(S_i(t))$ is the fitness of the solution Si and

$$\sum_{j=1}^{M} f(s_j(t))$$

is the total sum of all the members of the population, then the probability that the solution Si will be copied to the next generation is

$$\frac{f(s_i(t))}{\sum_{j=1}^{M} f(s_j(t))}$$

Another method for selecting the solution to be copied is tournament selection. Typically, in this method, the genetic program chooses two solutions randomly. The solution with the higher fitness will win. This method simulates biological mating patterns in which two members of the same sex compete to mate with a third one of a different sex. Finally, the third method is done by rank. In rank selection, selection is based on the rank (not the numerical value) of the fitness values of the solutions of the population. The creation of the offspring from the crossover operation is accomplished by deleting the crossover fragment of the first parent and then inserting the crossover fragment of the second parent. The second offspring is produced in a symmetric manner.

5.3.3.6 Mutation

The mutation is another important feature of genetic programming. Two types of mutations are possible. In the first kind a function can only replace a function or a terminal can only replace a terminal. In the second kind an entire subtree can replace another subtree.

5.3.4 Data Representation

The need for a good representation in evolutionary computation, and in artificial intelligence more generally, is called the *representation problem*. Genetic programming has two forms of representation: the variational and the generative. The variational representation is a static description of a program and is subject to evolutionary variation. The main requirement for a variational representation is evolvability: the evolution of programs of increasing fitness on a generational basis when subjected to genetic variation. The generative representation is a product of the variational representation and describes the dynamic form of a program. Its main requirement is that it can be executed. Yet, despite the different requirements of variational and generative representations, most GP systems do not distinguish between the two.

5.3.4.1 Biological Representations

Biology does distinguish between variational and generative representations. They are called, respectively, the genetic and the phenotypic. The genetic representation, from a reductionist viewpoint, is a linear, spatially distributed, sequence of heritable attributes. Each heritable attribute describes the amino acid sequence of a protein.

Development interprets these descriptions and generates proteins, the fundamental components of the phenotypic representation.

A group of proteins working upon a common task is called a biochemical pathway. The tasks carried out by biochemical pathways fall into three broad classes: metabolic, signaling, and gene expression. Of these, metabolic pathways are considered the most fundamental for they implement the pro-cessing behaviors of the cell, while signaling and gene expression pathways take on a configurational role.

Biochemical processing amounts to the manipulation of a cell's chemical state through systems of chemical reactions. Metabolic pathways are composed of enzymes, a group of proteins that carry out catalytic behaviors and enable reactions that would otherwise not be possible in the relatively low cellular temperatures. Enzymes achieve their catalytic behavior by binding to specific chemicals, the enzyme's substrates, and guiding their reaction. Cooperation within metabolic pathways emerges from product–substrate sharing between enzymes, where the product of one enzyme becomes the substrate of another.

5.3.4.2 Biomimetic Representations

Biological representations possess a number of qualities conceivably useful to, but not usually found, in genetic programming representations. These, include: the specialization of evolutionary and executable forms; evolvable representations, 'designed' for evolution; neutrality, increasing genetic diver-sity, and adaptability; less constrained behavior, giving more freedom to evolution; and positional independence, not limiting gene function to gene position. A number of GP systems mimic the genetic representation of biol-ogy. Many of these have introduced a developmental stage, allowing the genetic representation to be independent of the executable representation. This has been shown to increase genetic diversity and encourage neutrality. A number of these approaches also allow positionally independent genic units within the genome.

The mimicry of phenotypic representations is less common. However, computational idioms have been used to describe the action of enzymes and biochemical pathways. Analogs of enzyme activity have been used for com-putational purposes in the artificial domain. Evolutionary models of path-way development have also been attempted.

5.3.4.3 Enzyme Genetic Programming Representation

Enzyme genetic programming is a system that mimics biology in both genetic and phenotypic representations. Phenotypic representation is based upon an abstraction of metabolic pathways. The aim of the system is to evolve analogs of metabolic pathways within combinational logic circuits.

5.3.5 Summary

GP is used to model and control multi-level processes and to dictate their behavior according to fitness based upon automatically generated algo-rithms. The use of GP is going to be a considerable benefit in the future to the growing field of social and behavioral simulations.

Bibliography

Banzhaf, W., Nordin, P., Keller, R.E., and Francone, F.D. 1998. *Genetic Programming: An Introduction: On the Automatic Evolution of Computer Programs and Its Applications,* Morgan Kaufmann.

Brameier, M. and Banzhaf, W. 2007. *Linear Genetic Programming,* Springer, New York.

Cramer, N.L. 1985. A representation for the Adaptive Generation of Simple Sequential Programs, in Proceedings of an International Conference on Genetic Algorithms and the Applications, J.J. Grefenstette, ed., Carnegie Mellon University.

Crosby, J.L. 1973. *Computer Simulation in Genetics,* John Wiley & Sons, London.

Fogel, D.B., ed. 1998. *Evolutionary Computation: The Fossil Record,* IEEE Press, New York.

Fogel, D.B. 2000. *Evolutionary Computation: Towards a New Philosophy of Machine Intelligence,* IEEE Press, New York.

Goldberg, D.E. and Samtani, M.P. 1986. Engineering optimization via genetic algorithm, Proceedings 9th Conference on Electronic Computation, ASCE, pp. 471–482

Holland, J.H. 1975. *Adaptation in Natural and Artificial Systems,* University of Michigan Press, Ann Abor, MI.

Koza, J.R. 1992. *Genetic Programming: On the Programming of Computers by Means of Natural Selection,* MIT Press.

Ravindran, A., Ragsdell, K.M. and Reklaitis, G.V. 2006. *Engineering Optimization: Methods and Applications,* John Wiley & Sons Inc, Hoboken, New Jersey.

Shaw, D., Miles, J., and Gray, A. 2004. Genetic programming within civil engineering, Organization of the Adaptive Computing in Design and Manufacture Conference, IEEE, New York.

Sivanandam, S.N. and Deepa, S.N. 2008. *Introduction to Genetic Algorithms,* Springer, Berlin, Heidelberg, Germany.

Suttasupa, Y., Rungraungsilp, S., Pinyopan, S., Wungchusunti, P., and Chongstitvatana, P. 2011. A comparative study of linear encoding in genetic programming, Ninth International Conference on ICT and Knowledge. ISBN 978-1-4577-2162-5/11.

6

Hybrid Systems

6.1 Introduction

A hybrid intelligent system is one that combines at least two intelligent technologies. For example, combining a neural network with a fuzzy system results in a hybrid neuro-fuzzy system. The combination of probabilistic reasoning, fuzzy logic, neural networks, and evolutionary computation forms the core of soft computing, an emerging approach to building hybrid intelligent systems capable of reasoning and learning in an uncertain and imprecise environment.

Hybrid architectures can be of different models and stand-alone models. It may be independent, with non-interacting components which allow comparison of the two transformational models. The hybrid systems begin as one type (e.g., an artificial neural network [ANN]) and end up as another (e.g., a fuzzy logic network [FL]). Also, there are loosely coupled models which may decompose into separate ANN and FL components. The output of one passed to the other through data files as inputs.

The tightly coupled models are similar to loosely coupled systems with communication through memory instead of files. Components of fully integrated models share data structure, knowledge representation through rule base, and pattern matching to various type of problems.

6.1.1 Neural Expert Systems

Expert systems rely on logical inferences and decision trees and focus on modeling human reasoning. Neural networks rely on parallel data processing and focus on modeling a human brain. Expert systems treat the brain as a black box. Neural networks look at the brain's structure and functions, particularly at its ability to learn. Knowledge in a rule-based expert system is represented by IF–THEN production rules. Knowledge in neural networks is stored as synaptic weights between neurons.

In expert systems, knowledge can be divided into individual rules, and the user can see and understand the piece of knowledge applied by the system. In neural networks, one cannot select a single synaptic weight as a discrete piece of knowledge. Here knowledge is embedded in the entire network; it cannot be broken into individual pieces, and any changes of a synaptic weight may lead to unpredictable results. A neural network is, in fact, a black box for its user. If we combine the advantages of expert systems and neural networks, we can create a more powerful and effective expert system.

A hybrid system that combines a neural network and a rule-based expert system is called a neural expert system (or a connectionist expert system). Figure 6.1 shows the basic structure of a neural expert system.

The heart of a neural expert system is the inference engine. It controls the information flow in the system and initiates inference over the neural knowledge base. A neural inference engine also ensures approximate reasoning.

6.1.2 Approximate Reasoning

In a rule-based expert system, the inference engine compares the condition part of each rule with data given in the database. When the IF part of the rule matches the data in the database, the rule is fired and its THEN part is executed. The precise matching is required (inference engines cannot cope with noisy or incomplete data).

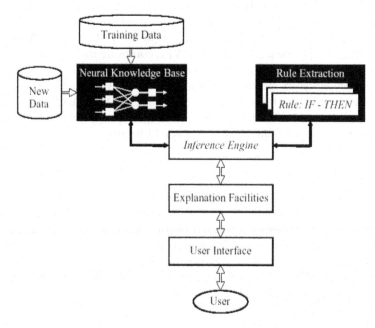

FIGURE 6.1
Basic structure of a neural expert system.

Neural expert systems use a trained neural network in place of the knowledge base. The input data does not have to precisely match the data that was used in network training. This ability is called *approximate reasoning*.

6.1.3 Rule Extraction

Neurons in the network are connected by links, each of which has a numerical weight attached to it. The weights in a trained neural network determine the strength or importance of the associated neuron inputs.

6.2 Neuro-Fuzzy

6.2.1 Neuro-Fuzzy Systems

Fuzzy logic and neural networks are natural complementary tools in building intelligent systems. While neural networks are low-level computational structures that perform well when dealing with raw data, fuzzy logic deals with reasoning on a higher level, using linguistic information acquired from domain experts. However, fuzzy systems lack the ability to learn and cannot adjust themselves to new environments. On the other hand, although neural networks can learn, they are opaque to the user.

Integrated neuro-fuzzy systems can combine the parallel computation and learning abilities of neural networks with the human-like knowledge representation and explanation abilities of fuzzy systems. As a result, neural networks become more transparent, while fuzzy systems become capable of learning.

A neuro-fuzzy system is a neural network which is functionally equivalent to a fuzzy inference model. It can be trained to develop IF–THEN fuzzy rules and determine membership functions for input and output variables of the system. The structure of a neuro-fuzzy system is similar to a multi-layer neural network. In general, a neuro-fuzzy system has input and output layers and three hidden layers that represent membership functions and fuzzy rules. Figure 6.2 shows a neuro-fuzzy system.

Each layer in the neuro-fuzzy system is associated with a particular step in the fuzzy inference process.

Layer 1 is the input layer. Each neuron in this layer transmits external crisp signals directly to the next layer. That is,

$$y_i^{(1)} = x_i^{(1)} \tag{6.1}$$

Layer 2 is the fuzzification layer. Neurons in this layer represent fuzzy sets used in the antecedents of fuzzy rules. A fuzzification neuron receives a crisp input and determines the degree to which this input belongs to the neuron's fuzzy set.

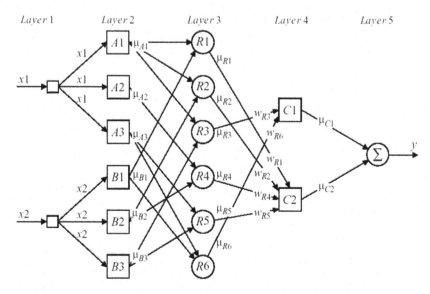

FIGURE 6.2
Neuro-fuzzy system.

Layer 3 is the fuzzy rule layer. Each neuron in this layer corresponds to a single fuzzy rule. A fuzzy rule neuron receives inputs from the fuzzification neurons that represent fuzzy sets in the rule antecedents.

Layer 4 is the output membership layer. Neurons in this layer represent fuzzy sets used in the consequent of fuzzy rules.

Layer 5 is the defuzzification layer. Each neuron in this layer represents a single output of the neuro-fuzzy system. It takes the output fuzzy sets clipped by the respective integrated firing strengths and combines them into a single fuzzy set.

6.2.2 Learning the Neuro-Fuzzy System

A neuro-fuzzy system is essential to a multi-layer neural network, and, thus, it can apply standard learning algorithms developed for neural networks including the back-propagation algorithm. When a training input-output example is presented to the system, the back-propagation algorithm computes the system output and compares it with the desired output of the training examples. The error is propagated backward through the network from the output layer to the input layer. The neuron activation functions are modified as the error is propagated. To determine the necessary modifications, the back-propagation algorithm differentiates the activation functions of the neurons.

6.2.3 Summary

The combination of fuzzy logic and neural networks constitutes a powerful means for designing intelligent systems. Domain knowledge can be put into a neuro-fuzzy system by human experts in the form of linguistic variables and fuzzy rules. When a representative set of examples is available, a neuro-fuzzy system can automatically transform it into a robust set of fuzzy IF–THEN rules, and thereby reduce our dependence on expert knowledge when building intelligent systems.

6.3 Neuro Genetic

6.3.1 Neuro-Genetic (NGA) Approach

In this approach, a genetic algorithm is used for updating the weight during the learning phase. A neural network with a configuration of 'l–m–n' has been considered for estimation, i.e., the network consists of 'l' number of input neurons, 'm' number of hidden neurons, and 'n' number of output neurons. For the input layer, the linear activation function is used, and for the hidden layer and output layer, the sigmoidal function or squashed-S function is used.

The number of weights N required for this network with a configuration of 'l–m–n' can be computed using the following equation:

$$N = (1+n)\_m \tag{6.2}$$

with each weight (gene) being a real number and assuming the number of digits (gene length) in weights to be d. The length of the chromosome L is computed using the following equation:

$$L - N\_d = (1+n)\_m\_d \tag{6.3}$$

For determining the fitness value of each chromosome, weights are extracted from each chromosome.

6.4 Fuzzy Genetic

Recently, numerous papers and applications combining fuzzy logic (FL) and genetic algorithms (GAs) have become known, and there is an increasing interest in the integration of these two topics. The use of FL-based techniques for either improving GA behavior and modeling GA components, the results obtained are called fuzzy genetic algorithms (FGAs), and the application of GAs in various optimization and search problems involving fuzzy systems are called genetic fuzzy systems (GFS).

Fuzzy genetic algorithms: An FGA is a GA that uses fuzzy logic-based techniques or fuzzy tools to improve the GA behavior in modeling different GA components.

An FGA may be defined as an ordering sequence of instructions in which some of the instructions or algorithm components may be designed with fuzzy logic-based tools, such as fuzzy operators and fuzzy connectives, for designing genetic operators with different properties, fuzzy logic control systems for controlling the GA parameters according to some performance measures, fuzzy stop criteria, representation tasks, etc.

Fuzzy optimization: Fuzzy optimization deals with how to find the best point under some fuzzy goals and restrictions given as linguistic terms or fuzzy sets.

GAs are used for solving different fuzzy optimization problems. This is the case, for instance, with fuzzy flow shop scheduling problems, vehicle routing problems with fuzzy due-time, fuzzy mixed integer programming applied to resource distribution, interactive fuzzy satisfying method for multi-objective 0-1, fuzzy optimal reliability design problems, job shop scheduling problems with fuzzy processing times, fuzzy optimization of distribution networks, etc.

Fuzzy neural networks: Neural networks have been recognized as an important tool for constructing membership functions, operations on membership functions, fuzzy inference rules, and other context-dependent entities in fuzzy set theory.

On the other hand, attempts have been made to develop alternative neural networks, more attuned to the various procedures of approximate reasoning. These alternative neural networks are usually referred to as fuzzy neural networks. The following features, or some of them, distinguish fuzzy neural networks from their classical counterparts: inputs are fuzzy numbers, outputs are fuzzy numbers, weights are fuzzy numbers, weighted inputs of each neuron are not aggregated by summation, but by some other aggregation operation. A deviation from classical neural networks in any of these features requires a properly modified learning algorithm to be developed.

GAs are used for designing an overall good architecture of fuzzy neural networks and fuzzy neural networks, for determining an optimal set of link weight, for participating in hybrid learning algorithms, etc.

Genetic fuzzy rule-based control systems: Fuzzy rule-based systems have been shown to be an important tool for modeling complex systems in which, due to the complexity or the imprecision, classical tools are unsuccessful.

GAs have proven to be powerful tools for automating the definition of the knowledge base of a fuzzy controller, since adaptive control, learning, and self-organization may be considered in many cases as optimization or search processes. Their advantages have extended the use of GAs in the development of a wide range of approaches for designing fuzzy controllers over the last few years. In particular, the application to the design, learning, and tuning of fuzzy controllers has produced promising results. These various approaches are collectively called genetic fuzzy systems (GFSs). On the other hand, we also must understand the GFSs as the application of GAs to any fuzzy system being the fuzzy rule-based systems (a particular case although the most extended), and this is the reason for calling this area genetic fuzzy rule-based control systems (FRBSs).

6.4.1 Genetic Fuzzy Rule-Based Systems

The idea of a genetic FRBS is that of a genetic FRBS design process which incorporates genetic techniques to achieve the automatic generation or modification of its knowledge base (or a part of it). This generation or modification usually involves a tuning/learning process, and, consequently, this process plays a central role in GFSs. The objective of this tuning/learning process is optimization, i.e., maximizing or minimizing a certain function representing or describing the behavior of the system.

It is possible to define two different groups of optimization problems in FRBSs. The first group contains those problems where optimization only involves the behavior of the FRBS, while the second one refers to those problems where optimization involves the global behavior of the FRBS and an additional system. The first group contains problems such as modeling, classification, prediction, and, in general, identification problems. In this case, the optimization process searches for an FRBS that is able to reproduce the behavior of a certain target system. The most representative problem in the second group is control, where the objective is to add an FRBS to a controlled system in order to obtain a certain overall behavior. Next, following are some aspects of the Genetic FRBSs.

6.4.2 The Keys to the Tuning/Learning Process

Regardless of the kind of optimization problem, i.e., given a system to be modeled/controlled (hereafter we use this notation), the involved tuning/learning process will be based on evolution. Three points are the keys to an evolutionary-based tuning/learning process:

1. The population of potential solutions
2. The set of evolution operators
3. The performance index

The population of potential solutions: The learning process works on a population of potential solutions to the problem. In this case, the potential solution is an FRBS. From this point of view, the learning process will work on a population of FRBSs, but considering that all the systems use an identical processing structure, the individuals in the population will be reduced to DB/RB or KB (knowledge-based). In some cases, the process starts off with an initial population obtained from available knowledge, whereas in other cases the initial population is randomly generated.

The set of evolution operators: The second question is the definition of a set of evolution operators that search for new and/or better potential solutions (KBs). The search reveals two different aspects: the exploitation of the best solution and the exploration of the search space. The success of evolutionary learning is specifically related to obtaining an adequate balance between exploration and exploitation, which finally depends on the selected set of evolution operators. The new potential solutions are obtained by applying the evolution operators to the members of the population of knowledge bases; each one of these members is referred to as an individual in the population. There are three evolution operators that work with a code (called a chromosome) representing the KB: selection, crossover, and mutation. Since these evolution operators work in a coded representation of the KBs, a certain compatibility between the operators and the structure of the chromosomes is required. This compatibility is stated in two different ways: work with chromosomes coded as binary strings (adapting the problem solutions to binary code) using a set of classical genetic operators, or adapt the operators to obtain compatible evolution operators using chromosomes with a non-binary code. Consequently, the question of defining a set of evolution operators involves defining a compatible couple of evolution operators and chromosome coding.

The performance index: Finally, the third question is that of designing an evaluation system capable of generating an appropriate performance index related to each individual in the population in such a way that a better solution will obtain a higher performance index. This performance index will drive the optimization process.

6.4.3 Tuning the Membership Functions

Another element of the KB is the set of membership functions. This is a second point where GAs could be applied with a tuning purpose. As in the previous case of scaling functions, the main idea is the definition of parameterized functions and the subsequent adaptation of parameters. The various proposals differ in the coding scheme and the management of the solutions (e.g., fitness functions).

6.4.4 Shape of the Membership Functions

Two main groups of parameterized membership functions have been proposed and applied: piecewise linear functions and differentiable functions.

6.4.5 The Approximate Genetic Tuning Process

As mentioned earlier, each chromosome forming the genetic population will encode a complete KB. More concretely, all of them encode the RB, R, and the difference between them are the fuzzy rule membership functions (i.e., the DB definition).

Taking into account a parametric representation with triangular-shaped membership functions based on a 3-tuple of real values, each rule

Ri: IF x1 is Ai1 and ... and xn is Ain THEN y is Bi,

of a certain KB (KBl), is encoded in a piece of chromosome Cli:

Cli = (ai1; bi1; ci1; : : : ; ain; bin; cin; ai; bi; ci)

where Aij, Bi have the parametric representation (aij ; bij ; cij), (ai; bi; ci), $i = 1; : : : : ;m$ (m represents

the number of rules), $j = 1; : : : : ; n$ (n is the number of input variables).

Therefore, the complete RB with its associated DB is represented by a complete chromosome.

Cl:

Cl = Cl1 Cl2 ::: Clm

This chromosome may be a binary or a real coded individual.

6.5 Summary

In this section, the hybrid approach among various soft computing techniques has been discussed. The limitations of every soft computing technique can be complemented by other techniques, so that the final outcome can be improved or make the approach more efficient.

Bibliography

Jang, J.S.R., Sun, C.T. and Mizutan, E. 1996. *Neuro-Fuzzy and Soft Computing: a Computational Approach to Learning and Machine Intelligence*, Prentice Hall, chapters 17–21, pp. 453–567.

Karrey, F. and de Silva, C. 2004. *Soft Computing and Intelligent System Design*, Addison Weley, chapter 7, pp. 337–361.

Rajasekharan, S. and Vijaylakshmi Pai, G.A. 2005. *Neural Network, Fuzzy Logic and Genetic Algorithm-Synthesis and Applications*, Prentice Hall, chapters 10–15, pp. 297–435.

7

Data Statistics and Analytics

7.1 Introduction

Data are indicators of information. The main purpose of data analysis is to derive usable and useful information. The analysis is irrespective of whether the data is qualitative or quantitative. Data analysis has multiple dimensions and approaches, bounded by diverse techniques under a variety of names, in different business, science, and social science domains. Describe and summarize the data, identify relationships between variables, compare variables and identify the difference between variables, and finally forecast outcomes. Most research problems are unlikely to be simply quantitative and qualitative but rather a mixture of these two approaches. There are many ways to analyze qualitative data, such as conducting interviews, transcription, and organization of data. These are the first stages of any data analysis. The interpretation of data will then be conducted by grouping together similar sets, systematically analyzing the transcripts, and, finally, drawing conclusions. Today, many computer-based technologies are available to deal with data analysis in faster and more appropriate ways. Figure 7.1 outlines the steps involved in the data processing.

The process of data analysis refers to separately examining each component in the data and converting it into useful information for decision-makers. Data requirements should be identified formerly in any project. Data collection can be done through downloading from online sources, satellites, questionnaires, etc. Data cleaning is a process of preventing and clearing incomplete or duplicate data present in the set. Mathematical equations and algorithms used to process the relationship between the variables present in data sets are called correlation causation. Once the analysis is completed, the next task is to display the results according to user requirements.

7.2 Data Analysis: Spatial and Temporal

A temporal data analysis is a technique used to examine and model the behavior of a variable in a data set over a time series. This behavior can be modeled as a function of previous data points of the same series, with

Data Science Process

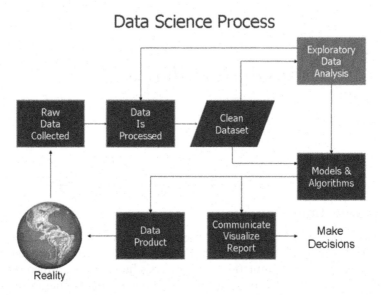

FIGURE 7.1
Steps involved in a data processing.

or without extraneous, random influences. These data are very common in the real world; they represent large sets of interactions between them and scales of variability. They can be applied to predictions in space and time, i.e., interpolation and forecasting, assimilation of observations and mechanistic models and inference on controlling process parameters. Common temporal analysis methods are time series plots, one-way ANOVA, sample autocorrelation, the rank von Neumann test, seasonality correlational, and the seasonal Mann–Kendall test.

7.2.1 Time Series Analysis

Time series analysis is a statistical technique that represents time-series data or a trend analysis. This will represent the raw data in a graphical view. This method is used to analyze time-series data in order to extract meaningful information. Time is represented by the X-axis and the data series observations are represented by the Y-axis. 'Time series' means that the data is from a series of particular time periods or intervals. Time series, cross sectional, and pooled data are the types of data considered for the analysis. In time-series data, the data is from observations of a variable at different periods of time. Cross-sectional data represents multiple variables collected at the same point in time. Pooled data is a combination of time series and cross-sectional data. Time series forecasting is applied to predict future values based on values observed in the past. An interrupted time series is the analysis of interventions on a single time series. They are different from spatial data analysis where the observations typically relate to geographical locations.

A stochastic model in a time series is more accurate when the observations are closer than when they are further apart. A time-series analysis can be applied to real-valued, continuous, and discrete numeric data. There are two solution techniques for time series: frequency domain and time-domain methods. Spectral and wavelet methods are the methods of solution followed in a frequency domain technique, whereas autocorrelation and cross-correlation are applied to the domain method. Figure 7.2 shows time series analysis with random data and a best-fit line.

7.2.2 One-Way ANOVA

One-way analysis of variance (one-way ANOVA) is a technique used to compare two or more samples; i.e., to check whether there are any statistically significant differences between the means of three or more independent groups. The following assumptions are made when working with one-way ANOVA:

- Response variable residuals are normally distributed.
- Variance of populations are equal.
- Responses for a given group are independent and identically distributed, normal random variables.

This is a procedure for testing the hypothesis that K population means are equal, where K > 2. The one-way ANOVA compares the means of the samples or groups in order to make inferences about the population means. It is also called a single factor analysis of variance because there will be only

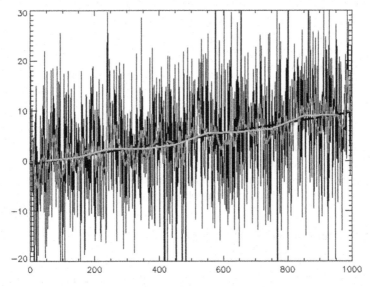

FIGURE 7.2
A time series analysis with random data and a best-fit line.

one independent variable or factor. In an ANOVA, there are two kinds of variables, namely, independent and dependent. The independent variable is controlled by the researcher. It is a categorical or discrete variable used to form the groupings of observations. In the one-way ANOVA, whereas only one independent variable is considered, two or more (theoretically, any finite number) levels of the independent variables are possible. The independent variable is typically a categorical variable. The independent variable (or factor) divides individuals into two or more groups or levels. There are two types of independent variables: active and attribute. If the independent variable is an active variable then we have to manipulate the values of the variable to study its effect on another variable; whereas, for an attribute independent variable, we do not have to alter the variable during the study.

7.2.3 Sample Autocorrelation

Autocorrelation refers to the correlation of a time series with its own past and future values. Autocorrelation is sometimes called 'serial correlation,' which refers to the correlation between members of a series with numbers arranged in time. Autocorrelation is just one measure of randomness. It complicates the application of statistical tests by reducing the effective sample size and the identification of significant covariance or correlation between time series. The randomness is ascertained by computing autocorrelation for data values at varying time lags. If random, such autocorrelations should be near zero for any and all time-lag separations. If non-random, then one or more of the autocorrelations will be significantly non-zero. Checking for autocorrelation is typically a sufficient test of randomness as the residuals from poor fitting models tend to display non-subtle randomness. But some applications require a more rigorous determination of randomness.

Time-series plot, lagged scatterplot, and autocorrelation functions are the three tools of sample autocorrelation. Figure 7.3 shows the plot for the autocorrelation function plot.

7.2.4 Rank von Neumann (RVN) Test

The RVN performs extremely well as a test for serial correlation data. It is a good alternative for the non-parametric test and is commonly used today. When this is applied to a set of raw observation in a non-regression, RVN is distribution-free, or robust, because the null distribution stays the same regardless of the distribution of the observation.

7.2.5 Seasonal Mann–Kendall Test

The seasonal Kendall (SK) test is a non-parametric test that analyzes data for monotonic trends in seasonal data. It is the most popular trend test in environmental studies. The SK test is a special kind of Mann–Kendall (MK) test. If the data is not seasonal, we can use the MK test instead. The SK test

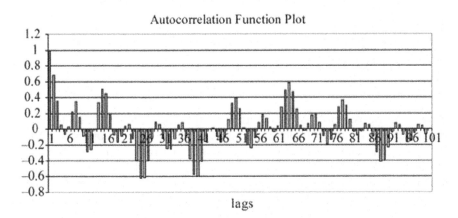

FIGURE 7.3
Autocorrelation function plot.

runs a separate MK trend test on each of m seasons separately, where m is the number of seasons. Data is only compared to the same season. For example, spring would be only be compared with spring, and summer would only be compared with summer. Non-parametric statistical procedures are commonly used to analyze time series for the trends in water resources.

7.3 Data Pre-Processing

Data processing is the conversion of data into a useful and desirable form. This can be performed either manually or automatically. Today, most of the data processing is done by using computers and digital data. Based on user requirements, the output can take the form of graphs, maps, tables, charts, etc. The processing of any data will naturally depend on the data processing techniques and software in use, as well as the complexity of the data.

Data pre-processing is a critical part of data mining, although it is often neglected. It includes cleaning, normalization, transformation, feature extraction, selection, etc. The product of pre-processed data is considered the final training set.

While dealing with real-world data, we might encounter incompleteness, noise, missing attributes, errors, inconsistent data, etc. Pre-processing is the most important process to be carried out in order to improve the quality of the outcome. The following steps of pre-processing should be performed before going forward:

1. Data cleaning
2. Data integration

3. Data transformation
4. Data reduction
5. Data discretization

7.3.1 Data Cleaning

Data cleaning is a process of identifying and correcting inaccurate or corrupt data from a set of tables or database. Cleaning is the process of identifying incomplete, inaccurate, or irrelevant parts of the data and replacing or deleting the wrong data. Incomplete or irrelevant data are very common when we are dealing with real sets of data. Some of the reasons this happens are: collection of data was performed with a malfunctioning instrument, human errors, computer errors, limitations in technology. Duplicate tuples are also required for the data cleaning. This is the process of cleaning the data by filling the missing values, smoothing the noisy data, identifying and removing the outliers, and resolving the inconsistencies. Dirty data will reduce the quality of any work. Therefore, data cleaning techniques should be employed before proceeding further. After cleaning, any data will become as consistent as the other data sets in the system. A common data cleaning process is the enhancement, where data is made more complete by adding related information. The steps of a data cleaning process are:

1. Fill the missing values; they may be an attribute or a class value.
   - Ignore the tuple: this step is usually done when a class label is missing.
   - Use the attribute mean to fill the missing value.
   - Use the attribute mean for all samples belonging to the same class.
   - Predict the missing value using a learning algorithm: handle the attribute with the missing value as a dependent variable and run a learning algorithm (usually Bayesian or decision tree) to predict the missing value.
2. Identify outliers and smooth out noisy data.
   - Binning: Sort the attribute values and partition them into bins (see 'Data Discretization' below); then smooth by bin means, bin median, or bin boundaries.
   - Sort the attribute values and partition them into bins (see "Unsupervised Discretization");
   - Then smooth by bin means, bin median, or bin boundaries.
   - Clustering: group values in clusters; then detect and remove outliers (automatically or manually).
   - Regression: smooth by fitting the data into regression functions.
3. Correct inconsistent data: use domain knowledge or expert decision.

7.3.2 Data Integration

Data integration is the process of combining data from the residing different sources and providing the user with a unified view of them. The process is particularly significant when dealing with a variety of stations which include both commercial and scientific domains. This has become crucial when, for instance two companies have decided to merge their data sets. The most well-known implementation of data integration is building an enterprise data warehouse. This enables a business to perform its analysis based on the data available in their data warehouse.

7.3.3 Data Transformation

Data transformation is the process of converting data from one format or structure to another, i.e., usually from the format of a source system into the required format of a new destination system. The usual process involves converting documents, but data conversions sometimes involve the conversion of a program from one computer language to another in order to enable the program to run on a different platform. The usual reason for this data migration is the adoption of a new system that's totally different from the previous one. The ease of data transformation is dependent on the type of changes required between a source system and a destination system. The tools and technologies required for the data transformation will depend on the size of data, complexity, format, and structure of the data involved. Data discovery, mapping, code generation, code execution, and data review are the steps included in a data transformation process and are based on the complexity, structure, and format of the data involved. In a data discovery process, data is profiled using profiling tools; sometimes it can be done using manually written profiling scripts to better understand the structure and characteristics of the data and to decide on the needs to be transformed. The data mapping refers to how the individual fields are mapped, modified, joined, filtered, aggregated, etc. To develop the desired quality of data, it is best to do a data mapping. In a code generation task, we can generate codes using SQL, Python, R, etc. to transform the data. In the next step, that code will be executed against the data to create the desired output. The complexity of the execution depends on the code generated by the developers. The final step is the data review process, which will check the output data on its transformation requirements. Data review, the final step in the process, focuses on ensuring the output data meets the transformation needs. The process is usually carried out by the end users. The following steps outline the transformation process:

1. Normalization:
   - Scaling attribute values to fall within a specified range.
     - Example: to transform V in [min, max] to V' in [0,1], apply $V' = (V - Min)/(Max - Min)$

- Scaling by using mean and standard deviation (useful when min and max are unknown or when there are outliers): $V' = (V - Mean)/StDev$

2. Aggregation: moving up in the concept hierarchy on numeric attributes.
3. Generalization: moving up in the concept hierarchy on nominal attributes.
4. Attribute construction: replacing or adding new attributes inferred by existing attributes.

7.3.4 Data Reduction

Complex data may require a very long time to process, which may not be feasible. Using data reduction techniques, we can simplify the analysis without compromising the integrity of the original data and produce a quality output. The concept of data analysis is to reduce the volume of the data sets.

These are common techniques used in data reduction:

- Order by some aspect of size.
- Rearrange rows and columns of tables to make patterns easier to see.
- Round drastically to one, or at most two, effective digits (effective digits are ones that vary in that part of the data).
- Use averages to provide a visual focus as well as a summary.
- Use layout and labeling to guide the eye.
- Remove the visual elements in a chart that are not necessary to comprehend the information represented on the chart.
- Give a brief verbal summary.

7.3.5 Data Discretization

Discretization of numerical data is one of the most influential data preprocessing tasks. It is a kind of data-reducing technique as it reduces the data from a large domain of numeric values to a subset of categorical values. Discretization causes remarkable improvement in the learning process. A typical discretization process generally consists of four steps:

- Sorting the continuous values of the feature to be discretized.
- Evaluating a cut-point for splitting or adjacent intervals for merging.
- Splitting or merging intervals of continuous values according to some defined criterion.
- Stopping at some point.

7.4 Presentation of Data

This is the most important task involving the visual representation of data. The primary goal of any data presentation task is to communicate the information very effectively and clearly using statistical graphs, plots, and information graphics. Information presentations are very powerful communication tools. They can make an article easy to understand, attract and sustain the interest of readers, and efficiently present large amounts of complex information. Inaccurate presentation will clearly fail to convince the end users. Numerical data may be encoded using dots, lines, or bars which are the effective way to communicate a quantitative message. Effective presentation helps users to analyze the problem in a very effective way. It makes complex data more accessible, understandable, and usable. Tables are generally used where users will look up a specific measurement, while charts of various types are used to show patterns or relationships in the data for one or more variables.

After the collection of data, it is necessary to summarize them suitably and present in such forms as can facilitate subsequent analysis and interpretation. There are three major techniques for presentation of data: tabular form, graphical form, and text form.

7.4.1 Tabular Presentation

Data may be presented in the form of statistical tables. In one table only simple frequencies can be shown. A table is best suited for representing individual information and represents both quantitative and qualitative information. Also, in the same table cumulative frequencies, relative frequencies, and cumulative relative frequencies can be shown. Relative frequencies and cumulative frequencies are defined as follows: relative frequency refers to the ratio of the frequency in the category of concern to the total frequency in the reference set; cumulative frequencies applies to an ordered set of observations from smallest to largest. The cumulative relative frequency is the sum of the relative frequencies for all values that are less than or equal to the given value.

7.4.2 Graphical Presentation

A graph is a very effective visual tool as it displays data at a glance, facilitates comparison, and can reveal trends and relationships within the data, such as changes over time, frequency distribution, and correlation or relative share of a whole. A graphical presentation should do the following:

- Show the data.
- Induce the viewer to think about the substance rather than about methodology, graphic design, or the technology of graphic production, etc.

- Avoid distorting what the data has to say.
- Present many numbers in a small space.
- Make large data sets coherent.
- Encourage the eye to compare different pieces of data.
- Reveal the data at several levels of detail, from a broad overview to the fine structure.
- Serve a reasonably clear purpose: description, exploration, tabulation, or decoration.
- Be closely integrated with the statistical and verbal descriptions of a data set.

There are many formats of graph to choose from: scatter plot, bar chart and histogram, pie chart, line plot, etc.

7.4.3 Text Presentation

Text is the main method of conveying information, as it is used to explain results and trends, as well as provide contextual information. Data are presented in paragraphs and sentences. Text can be used to provide interpretation or emphasize certain data. If quantitative information to be conveyed consists of one or two numbers, it is more appropriate to use written language than tables or graphs. If more data are to be presented, or other information such as that regarding data trends are to be conveyed, a table or a graph would be more appropriate. By nature, data take longer to read when presented as text; and when the main text includes a long list of information, readers and reviewers may have difficulty understanding the information.

7.5 Summary

In this section, various statistical inferences have been discussed. Testing of data, spatial and temporal analysis, and time-series data have been analyzed. The form of presentation such as tabular, graphical and text were introduced.

Bibliography

Agresti, A. and Finlay, B. 1997. *Statistical Methods for the Social Sciences*, 3rd Edn, Prentice Hall.
Anderson, T.W. and Sclove, S.L. 1974. *Introductory Statistical Analysis*, Houghton Mifflin Company.

Clarke, G.M. and Cooke, D. 1998. *A Basic Course in Statistics*, Arnold.

Electronic Statistics Textbook, http://www.statsoftinc.com/textbook/stathome.html.

Freund, J.E. 2001. *Modern Elementary Statistics*, Prentice Hall.

Johnson, R.A. and Bhattacharyya, G.K. 1992. *Statistics: Principles and Methods*, 2nd Edn, Wiley.

Moore, D. and McCabe, G. 1998. *Introduction to the Practice of Statistics*, 3rd Edn, Freeman.

Newbold, P. 1995. *Statistics for Business and Econometrics*, Prentice Hall.

Weiss, N.A. 1999. *Introductory Statistics*, Addison Wesley.

8

Applications in the Civil Engineering Domain

8.1 Introduction

The domain of civil engineering is a creative one. The problems encountered in this field are generally unstructured and imprecise, influenced by intuitions and past experiences of a designer. The conventional methods of computing that rely on analytical or empirical relations have become time-consuming and labor-intensive when posed with real-life problems. To study, model, and analyze such problems, approximate computer-based soft computing techniques inspired by the reasoning, intuition, consciousness, and wisdom possessed by human beings are employed. In contrast to conventional computing techniques which rely on exact solutions, soft computing aims to exploit a given tolerance of imprecision; the trivial and uncertain nature of the problem yields an approximate solution to a problem in quick time. Soft computing is a multi-disciplinary field, using a variety of statistical, probabilistic, and optimization tools which complement each other, such as neural networks, genetic algorithms, fuzzy logic, and support vector machines.

8.2 In the Domain of Water Resources

8.2.1 Groundwater Level Forecasting

Forecasting of groundwater levels is very useful for efficient planning of integrated management of groundwater and surface water resources in a basin. Accurate and reliable groundwater level forecasting models can help ensure the sustainable use of a watershed's aquifer for both urban and rural water supply. The present work investigates the potential of two neural networks, radial basis function neural networks (RBFNN) and generalized regression neural networks (GRNN), in comparison to regular ANN models, like feed forward back propagation (FFBP) and non-linear regression model (NARX) for modeling in groundwater level (GWL) forecasting of a coastal aquifer in western Ghats of India. A total of 24 wells (both shallow and deep) located within the study area (a micro watershed of Pavanje River Basin) were selected covering around 40 square km. Here, two different data sets, weekly time series GWL and meteorological variables, recorded during the study period (2004–11) were used in the analysis. Various performance indices such as root mean squared error (RMSE), coefficient of correlation (CC), and coefficient of efficiency (CE) were used as evaluation criteria to assess the performance of the developed models.

At the first stage, the potential and applicability of RBF for forecasting groundwater levels are investigated. Weekly time-series groundwater level data up to four lagged data has been used as various input scenarios where predicted outputs are one- and two-week lead time GWL. The analysis was carried out separately for three representative open wells. For all three well stations, higher accuracy and consistent forecasting performance of the RBF network model was obtained compared to the FFBP network model.

After confirming the suitability of RBF in GWL forecasting and with better accuracy over FFBP, the work has been extended further to consolidate the applicability of RBF in multi-step lead time forecasting up to six weeks ahead. In this study, six representative wells are covered for development of RBF models for six different input combinations using lagged time series data. Outputs are the predicted GWL up to six weeks. RBF models are developed for every well station and results are compared with nonlinear regression model (NARX). It has been observed that for all six well stations, the RBF network model achieved higher and more consistent forecasting, which consolidates the forecasting capability of RBF. The NARX model showed poor performance.

In the third stage, to examine the potential and applicability of GRNN in GWL forecasting, various GRNN models have been developed by considering the advantage of S-summation and D-summation layers for different input combinations using time series data. Weekly time-series groundwater level data up to four lagged data have been used as inputs where predicted outputs are one-week lead time GWL. The analysis has been carried out separately for three representative open wells. GRNN models were developed for every well and the best model results were compared with best RBF and FFBP using LM training algorithm models. The RBF and GRNN models performed similarly in GWL forecasting with higher accuracy in all the representative well stations. The poor performance of the FFBP-LM model was also satisfactory but found inferior to both GRNN and RBF.

After confirming the potential and applicability of GRNN and RBF in time series GWL forecasting with similar capability, the robustness, adaptability, and flexibility characteristics of these two techniques are further investigated for suitability with cause and effect relationship. Here, various meteorological parameters are used as causal variables and the GWL is used as output effect. Only GRNN models are developed in the present study as RBF was found with similar predicting performance in previous studies. Five various input combinations are used to obtain best results as one-step lead time output for three representative wells. In this case, the GRNN model predicted groundwater level with higher accuracy and with satisfactory results. The GRNN model performance was compared to a general ANN (FFBP) model and outperformed the FFBP.

The result of the study indicates the potential and suitability of RBFNN and GRNN modeling in GWL forecasting for multi-step lead time data. The performance of RBFNN and GRNN were almost equal. Although accuracy of forecasted GWL generally decreases with the increase of lead time, the GWL forecasts were within acceptable accuracy for both models. Figures 8.1–8.9 show the model performance. Table 8.1 presents the results.

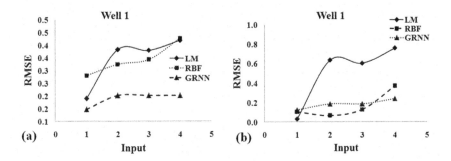

FIGURE 8.1
(a, b) Performance of different Modes for different inputs during training and testing at well no1.

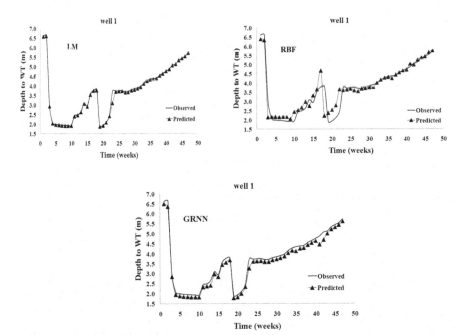

FIGURE 8.2
Performance of different Models in time series at well no1.

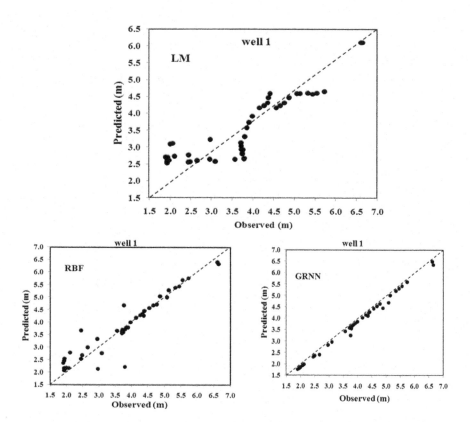

FIGURE 8.3
Scatter plots of different Models in time series at well no1.

Comparison between the LM, RBF and GRNN performance for well no. 2

Table 8.2 summarizes the predicted error for the developed models for the training and the testing period, showing that the ability to predict is much higher for GRNN.

Comparison between the LM, RBF, and GRNN Performance for Well No. 3

8.2.2 Water Consumption Modeling

In the present study, fuzzy models and hybrid models, such as ANFIS, consisting of fuzzy and neural networks, have been adopted to determine present and future water consumption of New Mangalore Port, India. In addition, multiple linear regression models were also developed for comparison purposes. The water consumption data is collected from New Mangalore Port for

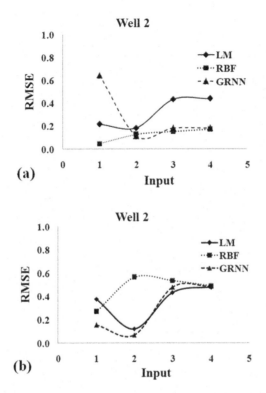

(a)

(b)

FIGURE 8.4
(a, b) Performance of different Modes for different inputs during training and testing at well no 2.

a period of six years and consist of different time steps such as daily, weekly, and monthly. The work is carried out in two stages, initially using time series data and, in the second stage, time-series data combined with meteorological data. In the first stage, three days previous and present day water consumption is used as inputs to determine the future water consumption. Similarly, in the second stage, previous day and present day water consumption with meteorological parameters is used as inputs. Various fuzzy and ANFIS models have been developed by changing different input scenarios, different membership functions, different fuzzy sets, different rules criteria, different defuzzification methods and using daily, weekly, and monthly data sets for single and multiple lead times. Also, the work carried on for different lengths of data sets, such as, all years, more than three years, and more than one year. Figures 8.10–8.12 show model performance. Table 8.5 shows model structure. Table 8.6, Table 8.7 and Table 8.8 present the various analysis of fuzzy models.

Table 8.9 presents model strategy.

Table 8.10 shows various rule formations.

Table 8.11 presents fuzzy logic model performance for various rules. Here, 12 rules performed better.

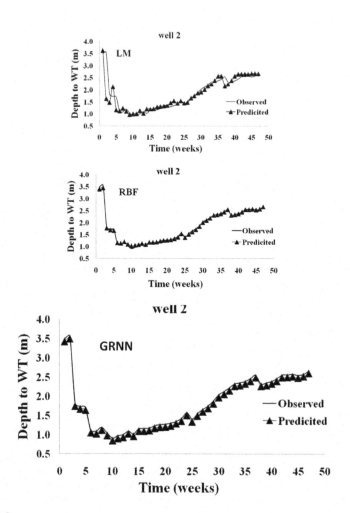

FIGURE 8.5
Performance of different Models in time series at well no 2.

Table 8.12 Presents comparison of fuzzy model performance using raw and normalized data. In most cases, the fuzzy logic model using normalized data performed better.

8.2.3 Modeling Failure Trend in Urban Water Distribution

Pipeline leakage is one of the crucial problems affecting urban water distribution systems from both an environmental and economic point of view. Unfortunately, the necessary large databases are not maintained in India for proper replacement of pipes. In this situation, this research purports using two artificial intelligence techniques, such as ANN and ANFIS to access the present condition and to predict the future trend of pipeline network of the

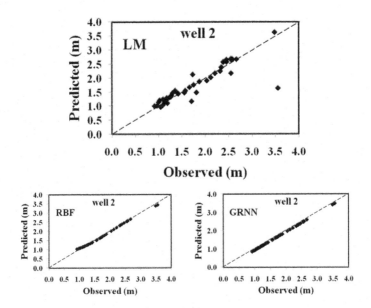

FIGURE 8.6
Scatter plots of different Models in time series at well no 2

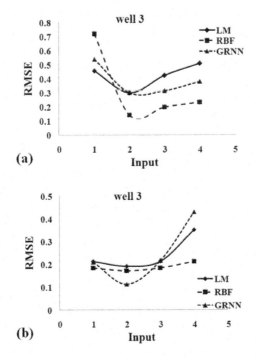

FIGURE 8.7
(a, b) Performance of different Modes for different inputs during training and testing at well no 3.

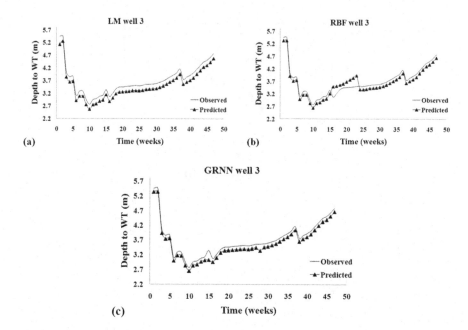

FIGURE 8.8
Performance of different Models in time series at well no 3.

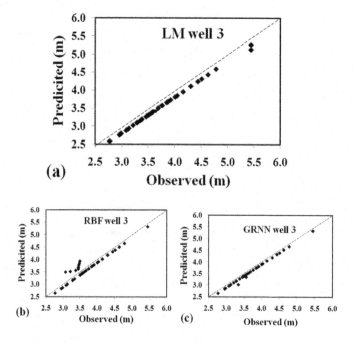

FIGURE 8.9
Scatter plots of different Models in time series at well no 3.

TABLE 8.1

Results of Training and Testing for Different Inputs Scenario for Well 1

Well 1			LM Training			LM Testing		
	Input	Hidden Neuron	RMSE	r	CE	RMSE	r	CE
LM	1	5	0.189951	0.990647	0.97664	0.640974	0.891067	0.751072
	2	20	0.383635	0.952539	0.904716	0.63777	0.873977	0.753554
	3	19	0.380125	0.957725	0.906452	0.603931	0.882685	0.779013
	4	17	0.418845	0.945593	0.886424	0.760757	0.812549	0.649341
RBF	1	3	0.280355	0.974862	0.949114	0.106487	0.859171	0.736964
	2	18	0.324377	0.965489	0.931879	0.065146	0.950562	0.901552
	3	20	0.343029	0.961196	0.92382	0.127429	0.790912	0.623334
	4	7	0.428129	0.939484	0.881333	0.028176	0.998253	0.981584
GRNN	1	17	0.147116	0.999633	0.985816	0.12375	1	0.990721
	2	7	0.201594	0.997645	0.973366	0.185625	1	0.979123
	3	9	0.201594	0.997645	0.973366	0.185625	1	0.979123
	4	10	0.201594	0.997645	0.973366	0.12375	1	0.990721

TABLE 8.2

Results of Training and Testing for Different Inputs Scenario for Well 2

Well 2			Training			Testing		
	Sl. No	Hidden Neuron	RMSE	r	CE	RMSE	r	CE
LM	1	28	0.218016	0.96886	0.933493	0.377635	0.875206	0.653833
	2	25	0.185781	0.975973	0.951706	0.121741	0.982014	0.964023
	3	21	0.435313	0.976165	0.73485	0.433574	0.973914	0.54368
	4	30	0.441448	0.932058	0.727324	0.481587	0.897713	0.437022
RBF	1	5	0.048272	0.999119	0.99674	0.269372	0.913026	0.823864
	2	11	0.128332	0.988575	0.976956	0.565826	0.620356	0.222843
	3	7	0.153763	0.983351	0.966918	0.537591	0.651405	0.298471
	4	11	0.171175	0.979289	0.959001	0.490642	0.861962	0.415651
GRNN	1	20	0.64625	1	0.410954	0.155031	1	0.946936
	2	10	0.114469	1	0.981519	0.06875	1	0.989565
	3	6	0.183047	1	0.952742	0.481877	0.777447	0.487337
	4	13	0.183047	1	0.952742	0.487169	0.768479	0.476014

Peoorkada zone in Trivandrum city, Kerala, India, where a huge amount is spent every year for leak rectification. Using different influential input variables, five models (all diameter pipes) have been developed. Also, the effect of each parameter (age, demand, and diameter) along with the previous year failures and previous year failures alone to current year failures are analyzed. Another three models of selective pipe diameter for considering

TABLE 8.3

Results of Training and Testing for Different Inputs Scenario for Well 3

Well 3			Training			Testing		
	Sl. No	Hidden Neuron	RMSE	r	CE	RMSE	r	CE
LM	1	19	0.4584	0.848398	0.692131	0.213785	1	0.864058
	2	10	0.30144	0.932349	0.866869	0.19305	1	0.889149
	3	6	0.424975	0.955056	0.735391	0.213785	1	0.864058
	4	6	0.509783	0.90187	0.619244	0.35321	1	0.628922
RBF	1	20	0.718135	0.995992	0.244404	0.186293	1	0.896773
	2	18	0.143167	0.985931	0.969969	0.173495	0.958473	0.910469
	3	11	0.199627	0.970489	0.941613	0.186293	1	0.896773
	4	14	0.23534	0.958613	0.918854	0.213916	0.99428	0.863891
GRNN	1	20	0.540507	0.929599	0.567451	0.210544	0.996427	0.868148
	2	5	0.302561	0.953533	0.864463	0.1144	1	0.961073
	3	17	0.315639	0.946477	0.852493	0.216848	0.996262	0.860135
	4	19	0.38145	0.950693	0.784569	0.431192	0.863298	0.44698

TABLE 8.4

Performance of LM, RBF, and GRNN Models during Training and Testing Periods

		Sl. No	Hidden Neuron	Training			Testing		
				RMSE	r	CE	RMSE	r	CE
Well 1	LM	1	5	0.189951	0.990647	0.97664	0.640974	0.891067	0.751072
	RBF	1	3	0.280355	0.974862	0.949114	0.106487	0.859171	0.736964
	GRNN	1	20	0.64625	1	0.410954	0.155031	1	0.946936
Well 2	LM	2	25	0.185781	0.975973	0.951706	0.121741	0.982014	0.964023
	RBF	1	5	0.048272	0.999119	0.99674	0.269372	0.913026	0.823864
	GRNN	1	20	0.64625	1	0.410954	0.155031	1	0.946936
Well 3	LM	2	10	0.30144	0.932349	0.866869	0.19305	1	0.889149
	RBF	2	18	0.143167	0.985931	0.969969	0.173495	0.958473	0.910469
	GRNN	2	5	0.302561	0.953533	0.864463	0.1144	1	0.961073

the influence of pre failures up to last year alone and pre failures up to last year along with length and pre failures up to last year along with length and demand to current year failures are constructed.

Prioritizing the pipeline replacement is done for mains having 400mm diameter and above, since network details pertaining to those diameters are available. The performance of the models is evaluated by using coefficient of correlation and mean absolute error and is compared to multiple linear regression (MLR) models. The three of them perform well and in kind, even though the ANN cascade forward backpropagation network is slightly

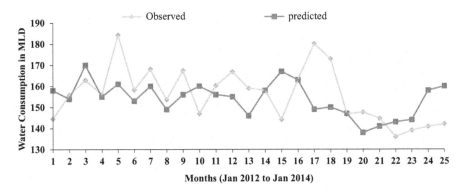

FIGURE 8.10
Observed and Estimated values of water consumption using fuzzy approach.

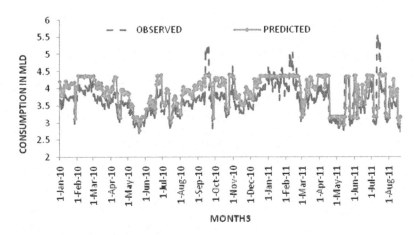

FIGURE 8.11
Time series plot of FL for more than four-year daily data set.

having an upper hand. The applicability and usefulness of ANN and ANFIS will surely help the authorities to make decisions regarding the replacement of pipes; this can, in turn, increase the efficiency of pipes.

8.2.4 Time Series Flow Forecasting

Recently ANN has shown great ability in modeling and forecasting nonlinear hydrologic time series. Although classic time series models like autoregressive moving average (ARMA) are widely used for hydrological time series forecasting, they are based on linear models assuming the data are stationary and have limited ability to capture non-stationarities and non-linearities in hydrologic data. ANNs are suitable for handling huge amounts of dynamic, nonlinear, and noisy data when underlying physical relationships

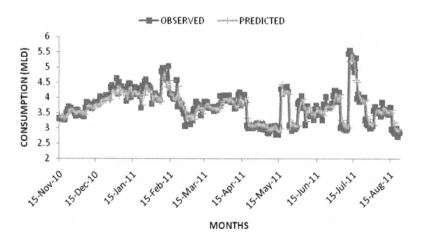

FIGURE 8.12
Time series plot of ANFIS for more than three-year data set.

TABLE 8.5

Model Structure

Data	Model	Inputs	Output
Time series	M1	WD(t)	WD(t+1)
	M2	WD(t)WD(t-1)	WD(t+1)
	M3	WD(t)WD(t-1)WD(t-2)	WD(t+1)
	M4	WD(t)WD(t-1)WD(t-2)WD(t-3)	WD(t+1)
Rainfall	M5	WD(t), RF(t)	WD(t+1)
	M6	WD(t) RF(t), WD(t-1)RF(t-1)	WD(t+1)
Temperature	M7	WD(t), Temp(t)	WD(t+1)
	M8	WD(t) temp(t), WD(t-1)Temp(t-1)	WD(t+1)
Relative Humidity	M9	WD(t), RH(t)	WD(t+1)
	M10	WD(t) RH(t), WD(t-1)RH(t-1)	WD(t+1)
Sunshine Hour	M11	WD(t), SH(t)	WD(t+1)
	M12	WD(t) SH(t), WD(t-1)SH(t-1)	WD(t+1)
Evaporation	M13	WD(t), EV(t)	WD(t+1)
	M14	WD(t) EV(t), WD(t-1)EV(t-1)	WD(t+1)

t-current time, t-1 previous time step, WD-water consumption, RH-relative humidity, SH-sunshine hours, EV-evaporation

are not fully understood. In spite of the high flexibility of ANN in modeling hydrologic time series, sometimes signals are highly non-stationary and exhibit seasonal irregularity. In such situations, ANN may not be able to cope with non-stationary data if preprocessing of input and/or output data is not performed. Simple ANN systems as well as complicated hybrid systems have been used to analyze real-world time series (hydrologic time series) which are usually characterized by mean and variance changes, seasonality, and other local behavior. Such real-world time series are not only invariably

TABLE 8.6

Rules Criteria Used in Fuzzy Logic Analysis

Rule No.	Rules Criteria
R1	IF inp1=mf1 AND inp2=mf1 AND inp3=mf1 THEN out1=mf1
R2	IF inp1=mf1 AND inp2=mf1 AND inp3=mf1 THEN out1=mf2
R3	IF inp1=mf1 AND inp2=mf1 AND inp3=mf1 THEN out1=mf3
R4	IF inp1=mf2 AND inp2=mf2 AND inp3=mf2 THEN out1=mf1
R5	IF inp1=mf2 AND inp2=mf2 AND inp3=mf2 THEN out1=mf2
R6	IF inp1=mf2 AND inp2=mf2 AND inp3=mf2 THEN out1=mf3
R7	IF inp1=mf3 AND inp2=mf3 AND inp3=mf3 THEN out1=mf1
R8	IF inp1=mf3 AND inp2=mf3 AND inp3=mf3 THEN out1=mf2
R9	IF inp1=mf3 AND inp2=mf3 AND inp3=mf3 THEN out1=mf3

inp is input, mf is membership function

TABLE 8.7

Fuzzy Set Used in Fuzzy Logic Analysis

No. of Fuzzy Sets	Types of Fuzzy Sets
Two	Low, High
Three	Low, Medium, High
Four	Low, Medium, High, Very High

TABLE 8.8

Fuzzy Set Used in Fuzzy Logic Analysis

Membership Function	Number of Fuzzy Set	Types of Fuzzy Set (Linguistic Variables)
Triangular	Three	Low, Medium, High
Trapezoidal	Four	Low, Medium, High, Very High

TABLE 8.9

Input and Output Model Combination

Sl. No.	Input	Output
Input 1	Rainfall (RF)	Water Consumption (WC)
Input 2	Maximum Temperature (T max)	
Input 3	Maximum Temperature (T max)	
Input 4	Relative Humidity (RH)	

nonlinear and non-stationary, but they also incorporate significant distortions due to both 'knowing and unknowing misreporting' and 'dramatic changes in variance.' The presence of these characteristics in time series stress the desirability of data preprocessing. These studies have focused on investigating the ability of NN to model non-stationary time series and the effect of data preprocessing on the forecast performance of NN.

TABLE 8.10

Rules Criteria Used in Fuzzy Model

No. of Rules	Different Rules Used to Develop the Models
R1	If (input1 is high) and (input2 is low) and (input3 is medium) and (input4 is high) then (output1 is medium)
R2	If (input1 is high) and (input2 is low) and (input3 is medium) and (input4 is high) then (output1 is high)
R3	If (input1 is medium) and (input2 is low) and (input3 is medium) and (input4 is medium) then (output1 is medium)
R4	If (input1 is medium) and (input2 is low) and (input3 is medium) and (input4 is medium) then (output1 is high)
R5	If (input1 is low) and (input2 is medium) and (input3 is high) and (input4 is medium) then (output1 is medium)
R6	If (input1 is low) and (input2 is medium) and (input3 is high) and (input4 is medium) then (output1 is high)
R7	If (input1 is low) and (input2 is medium) and (input3 is high) and (input4 is high) then (output1 is medium)
R8	If (input1 is low) and (input2 is medium) and (input3 is high) and (input4 is high) then (output1 is high)
R9	If (input1 is medium) and (input2 is low) and (input3 is medium) and (input4 is medium) then (output1 is low)
R10	If (input1 is high) and (input2 is low) and (input3 is medium) and (input4 is low) then (output1 is low)
R11	If (input1 is high) and (input2 is low) and (input3 is low) and (input4 is low) then (output1 is medium)
R12	If (input1 is high) and (input2 is medium) and (input3 is medium) and (input4 is high) then (output1 is high)

TABLE 8.11

Results of Fuzzy Model for Different Membership and Rules Criteria

Membership Type	Type	Three Rules	Six Rules	Nine Rules	Twelve Rules
Trapezoidal	RMSE (MLD)	14.69	25.08	15.27	12.38
	MAE (MLD)	4.24	7.24	4.40	3.57
Triangular	RMSE (MLD)	14.40	12.67	14.40	10.30
	MAE (MLD)	4.15	3.65	4.15	2.99

TABLE 8.12

Results of Fuzzy Model for Normalized Data

Model No.	Inputs	Output	Evaluation Type	Fuzzy Logic Method	
				Raw Data	Normalized Data
M1	RF	WC	PE (%)	37	24
			MAE (MLD)	73.74	47.91
			RMSE (MLD)	81.46	52.92
M2	Tmax	WC	PE (%)	28	22
			MAE (MLD)	56.82	45.32
			RMSE (MLD)	62.77	50.07
M3	Tmin	WC	PE (%)	22	30
			MAE (MLD)	45.07	59.82
			RMSE (MLD)	49.79	66.08
M4	RH	WC	PE (%)	24	20
			MAE (MLD)	49.24	41.16
			RMSE (MLD)	54.39	45.46
M5	RF, Tmax	WC	PE (%)	29	22
			MAE (MLD)	59.41	44.41
			RMSE (MLD)	65.62	49.05
M6	RF, Tmax, Tmin	WC	PE (%)	27	25
			MAE (MLD)	54.49	51.32
			RMSE (MLD)	60.19	56.69
M7	RF, Tmax, Tmin, RH	WC	PE (%)	27	25
			MAE (MLD)	54.49	51.32
			RMSE (MLD)	60.19	56.69
M8	Tmax, Tmin	WC	PE (%)	27	28
			MAE (MLD)	53.82	55.57
			RMSE (MLD)	59.45	61.39
M9	RF, RH	WC	PE (%)	27	23
			MAE (MLD)	54.49	46.16
			RMSE (MLD)	60.19	50.99
M10	Tmax, RH	WC	PE (%)	26	23
			MAE (MLD)	52.82	46.16
			RMSE (MLD)	58.38	50.99
M11	Tmin, RH	WC	PE (%)	24	26
			MAE (MLD)	48.24	53.16
			RMSE (MLD)	53.29	58.72
M12	RF, Tmin	WC	PE (%)	27	25
			MAE (MLD)	54.24	51.32
			RMSE (MLD)	59.92	56.69

TABLE 8.13

Statistical Characteristics of Daily Flow Data

Statistical Parameter	Pandu Station			Pancharatna Station		
	Training Set Q m³/s	Testing Set Q m³/s	All Data set Q m³/s	Training Set Q m³/s	Testing Set Q m³/s	All Data set Q m³/s
Minimum	2432	5539	2432	2086	1723	1723
Maximum	61015	54100	61015	75277	76236	76236
Mean	17520	18897	17904	16486	16550	16503
S_d	11011	10306	10836	13771	12086	13192
C_{xx}	0.59	0.64	0.59	0.99	0.81	0.95

8.2.5 Classification and Selection of Data

At Pandu Station, the data points are arranged for one-day, two-day, and three-day lagged data, which gives 6936 data points. Two-thirds of these are for the 'Training and Validation Data set'; and the remaining 1/3 are for the 'testing data set.' Thus we have used 4624 data points for training and validation and 2312 data points for testing. Similarly, for Pancharatna Station, 7302 data points are available, out of which 4868 data points are used for training and validation, and 2434 data points are for testing. The statistical characteristics of flow data are shown in Table 8.13 below, which reveals high variability of the flow data.

8.2.6 Overview of Research Methodology Adopted

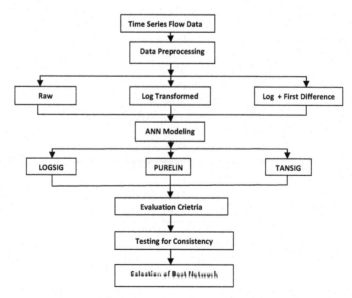

Three architectures of the ANN of the type multi-layer perceptron (MLP) are used. The Feed Forward Error Back Propagation (FFBP) algorithm is used. The MLP has three layers, viz., input layer, hidden layer, and output layer. The neurons of the hidden layer are varied from one to ten. The architectures PURELIN, LOGSIG, and TANSIG are used. These are according to the activating function of the ANN.

For each architecture, the number of neurons is varied from one to ten. This accounts for 30 types of networks. There are three types of data sets: raw, log transformed, and log plus first difference. In each data set, the number of lagged terms varies from one to three. Thus, there are nine inputs for each type of data set, making a total of 30 network types X 9 input data sets X 2 (Validation and Testing) = 540 trials at each station.

Selected tables below summarize the analysis at Pancharatna Station. The minimum errors which are consistent in both the data sets and where RMSE and MAPE are not in contradiction are marked by highlighting. The combined tables for the three data sets indicate best performance in each category. From this the final selection is shown by highlighting it. The results are also represented graphically below.

The results for Pandu Station also follow a similar trend, and the log transformed data set noticeably improves the network performance. Only the consolidated tables are presented herewith. The final selection of network architecture–data set combination is indicated by highlighting.

The results of raw data sets show much variation and are not consistent. These also show large errors. The log transformation shows improvement in performance to a marked degree, as well as consistency of results and stable performance of networks. The log + first difference decreases the accuracy of performance, but it does not show the inconsistency of the raw data set. The accuracy of prediction seems to increase when the number of inputs to

TABLE 8.14

Comparison of Network Performance for Testing Data Sets (Pancharatna)

Data Set	No. of Lagged Terms	Best Network Structure	LOGSIG		PURELIN		TANSIG	
			RMSE (m³/s)	MAPE (%)	RMSE (m³/s)	MAPE (%)	RMSE (m³/s)	MAPE (%)
Raw	1	1 – 5 – 1	1480.4	4.38	2845.6	29.7	1503.3	4.48
	2	2 – 7 – 1	1446.8	4.30	2841.0	29.9	22526	84.30
	3	3 – 10 – 1	1310.2	3.9	2831.1	29.6	1365.2	3.80
Log Transformed	1	1 – 3 – 1	1460.1	4.06	1736.3	5.98	1626.1	6.00
	2	2 – 10 – 1	1411.5	3.72	1697.8	5.87	1391.4	3.52
	3	3 – 5 – 1	1337.9	3.46	1673.2	5.88	1397.9	3.96
Log + First Difference	1	1 – 5 – 1	3010.5	7.56	3079.2	8.43	3073.7	8.48
	2	2 – 6 – 1	2929.1	7.40	2857.9	8.27	2525.1	7.01
	3	3 – 8 – 1	2630.8	6.83	2855.8	8.38	3487.3	8.57

TABLE 8.15

Comparison of Network Performance for Testing Data Sets (Pandu)

	No. of Lagged Terms	Best Network Structure	LOGSIG		PURELIN		TANSIG	
Data set			RMSE (m³/s)	MAPE (%)	RMSE (m³/s)	MAPE (%)	RMSE (m³/s)	MAPE (%)
Raw	1	1-4-1	1765.99	4.87	2279.11	9.53	1817.12	5.02
	2	2-7-1	1115.43	3.39	18116.07	75.91	18488.26	81.57
	3	3-2-1	1117.38	3.39	1671.52	8.63	19227.07	82.66
Log Transformed	1	1-4-1	1020.61	2.67	1473.88	4.50	1018.50	2.76
	2	2-4-1	909.22	2.17	1383.70	4.14	946.19	2.23
	3	3-6-1	907.95	2.12	1379.97	4.09	913.66	2.21
Log + First Difference	1	1-8-1	1812.85	4.27	1956.57	5.44	1816.48	4.37
	2	2-4-1	1923.52	4.71	1957.12	5.43	1818.27	4.30
	3	3-8-1	1774.85	4.34	1951.85	5.30	1863.13	4.49

the ANN is increased as seen by decreasing errors from 1-day lag data sets to 3-day lag data sets.

Graphs for comparison of the predicted and actual values for Pandu are shown in the Figures 8.13 and 8.14.

It is seen that the predicted streamflow values closely follow the pattern of actual values. Similar results are seen for the prediction at Pandu.

The consistency and stability of the selected networks is checked by swapping the training and testing data set selection. Table 8.16 shows statistical characteristics of swapped datasets and Table 8.17 shows model performance consistency test-swapped value.

- For the prediction of streamflow at Pandu Station, the ANN with LOGSIG activating function, log transformed data set, six neurons in the hidden layer is found most efficient, whereas at Pancharatna, a similar network with five neurons in the hidden layer is found most efficient.

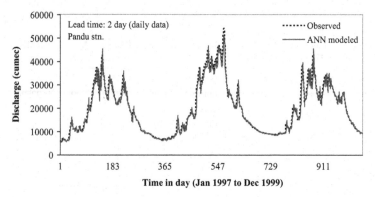

FIGURE 8.13

ANN model performance for 2 day ahead prediction.

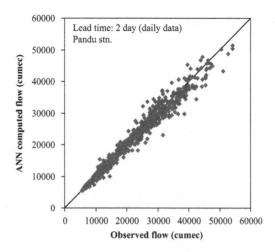

FIGURE 8.14
Scatter plot between observed and ANN modeled flow for 2-day lead time (daily data) for Pandu station during testing period.

TABLE 8.16

Statistical Characteristics of Swapped Data Sets

Discharge Value Q m³/s	Pandu Station			Pancharatna Station		
	Training Data Set	Testing Data Set	All Data Set	Training Data Set	Testing Data Set	All Data Set
Minimum	3008	2432	2432	1723	2086	1723
Maximum	61015	51319	61015	76236	72914	76236
Average	18545	16637	17904	16270	16984	16503

TABLE 8.17

Model Performance Consistency Test-Swapped Value

Station	Pancharatna				Pandu			
Error	RMSE TR	MAPE TR	RMSE TST	MAPE TST	RMSE TR	MAPE TR	RMSE TST	MAPE TST
BeforeSwap	1419.25	3.58	2005.90	5.15	1042.40	2.64	974.98	2.91
After Swap	1717.80	4.30	1337.90	3.46	1002.80	2.67	907.95	2.12

- Log transformation as a preprocessing technique for a time series of large variance and non-stationary nature is a very effective tool to improve the prediction by ANNs.

- For a nonlinear data set, the activation function PURELIN does not perform well; LOGSIG is most suitable for nonlinear data sets.

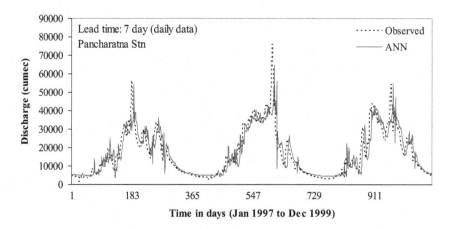

FIGURE 8.15
Time series plot between observed and ANN modeled flow for 7-day lead time (daily data) for Pancharatna station during testing period.

- The number of neurons in the hidden layer of the ANN affects the performance. The accuracy increases due to increasing the number up to a certain extent, only after which a plateau is reached and the performance remains unaffected (or may decrease in some cases if the number of neurons is further increased).

- Increasing the number of inputs improves the performance of the ANNs up to a certain extent, after which the increase in inputs may not have any effect on the performance.

- Incorporating the first difference of successive terms in the input for a time series analysis does not improve the accuracy of ANNs and, in fact, may decrease it.

Figure 8.13 shows ANN model prediction for 2-day lead time at Pandu. Figure 8.14 shows scatter plot. Figure 8.15 shows 7-day lead time prediction. Figure 8.16 shows scatter plot at Pancharatna.

8.3 In the Field of Geotechnical Engineering

The predictions of load capacity, particularly those based on pile-driving data, have been examined by several ANN researchers such as to predict the friction capacity of piles in clays. The neural network was trained with field data of actual case records. The model inputs were the pile length, the pile diameter, the mean effective stress, and the undrained shear strength. The skin friction resistance was the only model output. The methods were

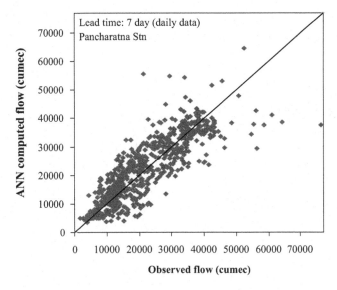

FIGURE 8.16
Scatter plot between observed and ANN modeled flow for 7-day lead time (daily data) for Pancharatna station during testing period.

compared using regression analysis as well as the error rate. It is evident that ANNs outperform the conventional methods. The study also pointed out that the main criticism of the ANN methodology is its inability to trace and explain the logic it uses to arrive at the outputs from the inputs.

It is evident from literature that ANNs have been applied successfully to many geotechnical engineering areas. This includes pile capacity prediction, settlement of foundations, soil properties and behavior, liquefaction, site characterization, earth retaining structures, slope stability, and the design of tunnels and underground openings. Perhaps the most successful and well-established applications are the capacity prediction of driven piles, liquefaction, and the prediction of soil properties and behavior. Other applications (e.g., settlement of structures) need to be treated with caution until additional research has been conducted. There are also several areas in which the feasibility of ANNs has yet to be tested, such as bearing capacity, prediction of shallow foundations, capacity of bored piles, design of sheet pile walls, and dewatering, among others.

Based on the results of the studies reviewed in this chapter, it is evident that ANNs perform better than, or as well as, the conventional methods used as a basis for comparison in many situations, whereas, they fail to perform well in a few.

In many situations in geotechnical engineering, it is possible to encounter some types of problems that are very complex and not well understood. For most mathematical models that attempt to solve such problems, the lack of physical understanding is usually supplemented by either simplifying the problem or incorporating several assumptions into the models. Mathematical models also rely on assuming the structure of the model in advance, which may be sub-optimal.

Consequently, many mathematical models fail to simulate the complex behavior of most geotechnical engineering problems. In contrast, ANNs are based on the data alone in which the model can be trained on input–output data pairs to determine the structure and parameters of the model. In this case, there is no need to simplify the problem or incorporate any assumptions. Moreover, ANNs can always be updated to obtain better results by presenting new training examples as new data become available.

Despite their good performance in many situations, ANNs suffer from a number of shortcomings, notably: the lack of theory to help with their development, the fact that success in finding a good solution is not always guaranteed, and their limited ability to explain the way they use the available information to arrive a solution. Consequently, there is a need to develop some guidelines, which can help in the design process of ANNs. There is also a need for more research to give a comprehensive explanation of how ANNs arrive at a prediction.

A further issue that needs to be given some attention in the future development of ANNs is the treatment of uncertainties associated with geotechnical engineering parameters. Despite these uncertainties, ANN models that have been developed to date in the field of geotechnical engineering are essentially deterministic. Consequently, procedures that incorporate such uncertainties into ANN models are essential, as they provide more realistic solutions. Overall, despite the limitations of ANNs, they have a number of significant benefits that make them a powerful and practical tool for solving many problems in the field of geotechnical engineering.

8.4 In the Field of Construction Engineering

Ready-mixed concrete (RMC) is manufactured in a factory or batching plant, according to a set recipe, and delivered to a work site by truck-mounted transit mixers. This results in a precise mixture, allowing specialty concrete mixtures to be developed and implemented on construction sites. A high amount of precision is required for the production of concrete of a specific grade. The various raw materials are weight batched and fed to the mixer through conveyors. All these operations are controlled by a computer housed in the control room.

RMC, as it is popularly called, refers to concrete that is specifically manufactured for delivery to the customer's construction site in a freshly mixed and plastic or unhardened state. Concrete itself is a mixture of Portland cement, water, and aggregates comprising sand and gravel or crushed stone.

At traditional work sites, each of these materials is procured separately and mixed in specified proportions on-site to make concrete. Today, RMC is usually bought and sold by volume (usually expressed in cubic meters) and can be custom-made to suit different applications. It is manufactured under computer-controlled operations and transported and placed on-site using sophisticated equipment and methods. Ready-mix concrete has cement, aggregates, water and other ingredients, which are weigh batched at a centrally located plant. This is then delivered to the construction site in transit mixers and can be used straight away without any further treatment.

The use of RMC is an environmentally friendly practice that ensures a cleaner work place and causes minimal disturbance to its surroundings. This makes its utility more significant in crowded cities and sensitive localities.

In contrast to this, conventional methods of making, transporting and placing concrete at most construction sites are somewhat labor-intensive and suffer from practices which may be erratic and not very systematic. Therefore, the use of ready-mixed concrete is more cost effective in the long term, while ensuring that structures are built faster and using concrete that comes with higher levels of quality assurance.

As observed, in the case of RMC manufacturing and delivery, the whole process has to cross a number of critical paths: plant efficiency, delivery system time management, travel conditions which are subject to high traffic congestion and environmental conditions, maintaining concrete quality, during production which might be affecting their performance in terms of concrete strength, delivery system etc.

In order to predict the accuracy of the compressive strength of concrete and the accuracy of mix proportions during preparation of concrete, a modeling study was conducted using an Artificial Neural Network study and compared to the results obtained.

A method to predict 28 days compressive strength of concrete by using neural networks was proposed based upon the inadequacy of present methods dealing with multiple variables and nonlinear problems. A model was built to implement the complex nonlinear relationship between the inputs (mix

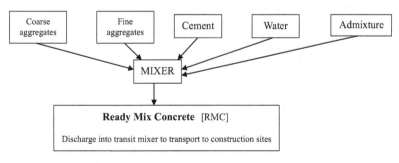

FIGURE 8.17
Working principle of RMC plants.

proportions) and the output (concrete strength). Another model was built to check the accuracy of the mix proportions used in the plant during the preparation of concrete with the design mix;. The mix proportions were used as the inputs, and the summation of these inputs were used as the output.

The neural network (NN) models showed high prediction accuracy and the research results conformed to some rules of mix proportion of concrete. These demonstrate that using neural networks to predict concrete strength is practical and beneficial. Thus, it was concluded that,

- Neural network model was constructed which provide a quick mean of predicting 28-day compressive strength of concrete based on mix proportion.
- The minimum accuracy of the network was 90%.
- This computational intelligent method will be helpful to civil engineers, technologists, ready-mix operators and concrete mixture designers in civil engineering and concrete mixing and batching plants.
- NN models attain good prediction accuracy. Some effects of concrete compositions on strength are in accordance with the rules of mix proportioning.
- Consequently, the application of NN models to concrete strength prediction is practical and has a good future.

The data used in the model development were taken from observations of the ongoing project in the construction of building for mechanical block inside the campus of National Institute of Technology, Surathkal, Karnataka. The project involved different contractors and different suppliers of RMC.

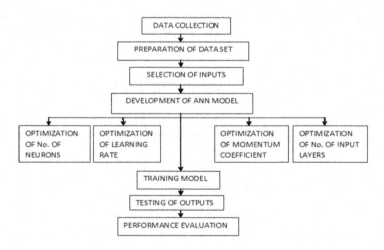

FIGURE 8.18
Proposed methodology.

These observations provided different examples of the ready-mix concrete supply, including the collection of all other possible parameters.

The data collected included: concrete mix proportions with targeted volume relating to mix design and actual volume according to the quantity mixed in the preparation of concrete in plant as per the batch report of individual truck mixes; quantity and type of concrete mix required at site; batch time; truck arrival time at site; travel time; wait time; discharge time; and average inter-arrival time. Data were also collected for the slump value (measure of consistency and fluidity) of concrete on-site and the compressive strength of concrete after 28 days. Several values, which were not available from the field, were assumed hypothetically. The table and graph below show the actual volume of 20mm provided to the targeted volume corresponding to the weight for individual mixes of concrete in cubic meters. The data were noted for all the materials used in the individual mixes such as 20mm, 12mm, sand, cement, water, and admixture. The percentage variation was calculated and the data were used as the inputs for the prediction of compressive strength and mix proportion.

TABLE 8.18

Sample of Data Set Collected from Field and Difference with Percentage Variation

Concrete	20 mm			%	Sand			%
In Cu.m	AV	TV	Diff.	Variation	AV	TV	Diff.	Variation
1	540	546	6	1.0989	697	711	14	1.969058
2	536	546	10	1.8315	695	711	16	2.250352
3	542	546	4	0.7326	701	711	10	1.40647
4	537	546	9	1.64835	712	711	−1	−0.14065
5	550	546	−4	−0.7326	714	711	−3	−0.42194
6	562	546	−16	−2.9304	731	711	−20	−2.81294
Total	3267	3276	−9	−0.2747	4250	4266	−16	−0.37506

AV- Actual value, TV- Targeted value

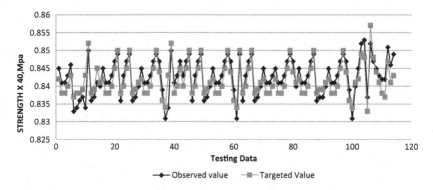

FIGURE 8.19
Comparison of model results and predicted results in testing process.

Different neural network models were developed for various input scenarios. The next stage was to select the optimum NN for each type of model setup (i.e., number of input layers, number of neurons, learning rate, and momentum coefficient) for use in the validation process. This selection was based upon the monitoring minimum values of MRE and MSE that were produced in the training and testing processes. Seeking a compromise in the training and testing MSE, so they are both low in relation to those of the other NN models developed, is a method of avoiding the problems of over-training and under-training. The MRE and MSE values are plotted with respect to the parameters used for optimization.

The training and testing results are then set for validation by finding the correlation coefficient for the various parameters during the optimization process. Correlation values are plotted with respect to the change in parameters. The optimum results were arrived at through trial and error.

Results and discussion are divided into two parts, prediction of compressive strength and prediction of accuracy in mix proportioning.

All the results obtained are presented in the Table 8.19. As observed from the table, the best model developed was for seven input parameters with neurons-5, learning rate-0.9 and momentum coefficient-0.4, considering the correlation coefficient which is the highest among all during both testing and training. The MSE and MRE values were also lowest for these combinations.

For various other input combinations, both MSE and MRE are of higher values and they are almost the same, as no significant difference was observed. The results are also graphically presented in the following section. Tables 8.20 and 8.21 present the influence of inputs and learning rate on error distribution.

TABLE 8.19

Comparison of Training and Testing Results between Specifications of Different Architectures

Number of Inputs	Number of Outputs	Number of Nuerons	Learning Rate	Momentum Coefficent	MRE	MSE	Cc (Train)	Cc (Test)
7	1	5	0.6	0.4	0.339855	0.000016	0.822	0.831
6	1	5	0.6	0.4	0.479885	0.000028	0.601	0.677
5	1	5	0.6	0.4	0.452766	0.000037	0.476	0.168
4	1	5	0.6	0.4	0.569415	0.000064	0.473	0.588
3	1	5	0.6	0.4	0.482263	0.000119	0.545	0.735
2	1	5	0.6	0.4	0.608362	0.000169	0.369	0.256
7	1	1	0.6	0.4	0.340148	0.000015	0.824	0.831
7	1	2	0.6	0.4	0.341195	0.000015	0.823	0.831
7	1	3	0.6	0.4	0.342065	0.000016	0.833	0.83
7	1	4	0.6	0.4	0.341058	0.000016	0.831	0.83
7	1	5	0.6	0.4	0.339855	0.000016	0.822	0.831

TABLE 8.20

Variation of MSE with Different Iterations for Various Number of Inputs

No. of Iterations	7 Inputs	6 Inputs	5 Inputs	4 Inputs	3 Inputs	2 Inputs
10000	0.000175	0.000176	0.000177	0.000177	0.000178	0.000178
25000	0.000155	0.000163	0.000171	0.000175	0.000178	0.000178
50000	0.000055	0.000062	0.00011	0.000145	0.000176	0.000177
75000	0.000025	0.000034	0.000053	0.000099	0.000167	0.000176
100000	0.000019	0.00003	0.00004	0.000071	0.000138	0.000174
120000	0.000016	0.000028	0.000037	0.000064	0.000119	0.000169

TABLE 8.21

Variation of MSE with Different Iterations for Various Learning Rates

No. of Iterations	L/R 0.1	L/R 0.2	L/R 0.3	L/R 0.4	L/R 0.5	L/R 0.6	L/R 0.7	L/R 0.8	L/R 0.9
10000	0.000178	0.000178	0.000177	0.000177	0.000177	0.000175	0.000172	0.00016	0.000098
25000	0.000177	0.000177	0.000176	0.000176	0.000169	0.000155	0.000113	0.000054	0.000027
50000	0.000172	0.000166	0.000153	0.000153	0.000083	0.000055	0.00003	0.000021	0.000011
75000	0.000137	0.000109	0.000079	0.000079	0.000034	0.000025	0.000019	0.000013	0.000008
100000	0.000077	0.000055	0.000038	0.000038	0.000023	0.000019	0.000014	0.00001	0.000007
120000	0.000048	0.000036	0.000028	0.000028	0.00002	0.000016	0.000012	0.000009	0.000007

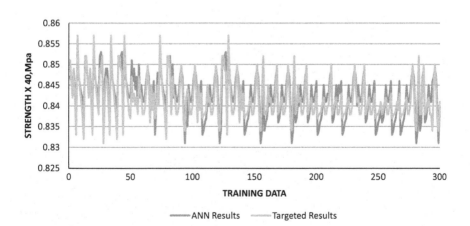

FIGURE 8.20

Comparison of BEST model results and predicted results in training process.

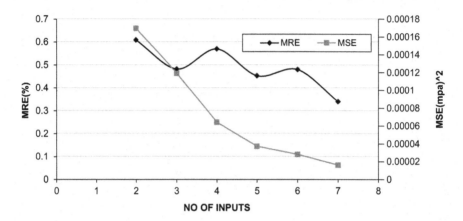

FIGURE 8.21
MSE and MRE with number of inputs.

8.4.1 Using Fuzzy Logic System: Methodology and Procedures

As was indicated in the first section, the objective of the study was to develop a fuzzy model for the prediction of compressive strength of ready-mixed concrete.

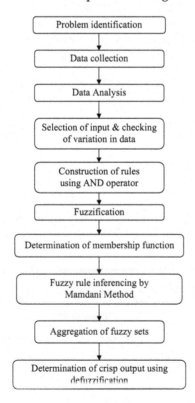

Selection of input variables: Here, the input variables are ingredients of the concrete, while the output response is compressive strength of the concrete. Considering the high fluctuation in the data series, out of seven input parameters, only three parameters (20mm aggregate, 12mm aggregate, and sand) are considered for fuzzification, as the other four parameters (cement1, cement2, water, and HRWRA-high range water reducing agent) are almost constant.

Checking of variation in data series: In the first part, before carrying out the strength analysis, volumetric analysis has been carried out. For this analysis, the inputs are the same as strength analysis. Volume of concrete is the output. Input parameters considered for the analysis are based on variation in the input data; if it is more than 3%–4% of the total quantity of each input, then it is considered for the analysis. If variation is less than 3% it is kept as constant. Input and output quantities used in the fuzzy model are given in Table 8.22.

Mamdani Fuzzy Inference System (FIS)

- **Construction of rules using AND operator**

    Set of rules were prepared. A single if–then rule is written as:

    IF "X" is A, THEN "Y" is B

    In the mathematical form;

    $\{$IF $(premise_{iy})$ THEN $(consequent_i)\}_{i=1}^{N}$

Where A and B are linguistic values defined by fuzzy sets on the ranges; X and Y, respectively. The if part of the rule "X is A" is called the antecedent or premise, while the then-part of the rule "Y is B" is called the consequent or conclusion. Some of the rules prepared are shown in Tables 8.23 and 8.24.

Some of the rules obtained for volumetric analysis are presented in Tables 8.23 and 8.24, and rules obtained from the strength analysis are presented in Tables 8.24 and 8.25.

Fuzzification: In fuzzification, considering minimum and maximum numerical value of the particular variable, the entire input domain is classified into three fuzzy or linguistic sets as Low, Medium and High. The classification range of the input and out parameters is presented in Table 8.25.

Considering 50% overlapping in the range value, each variable is classified as Low, Medium, and High.

Membership function: When the membership functions are defined, fuzzification takes a real-time input value. For the analysis, three different membership shapes are incorporated with fuzzification: triangular, trapezoidal and s-shape. Membership functions are

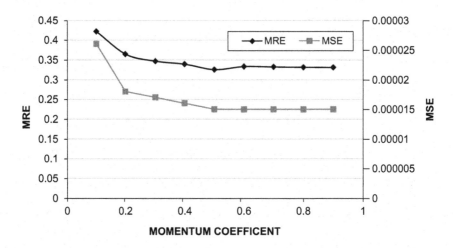

FIGURE 8.22
MSE and MRE with momentum coefficient.

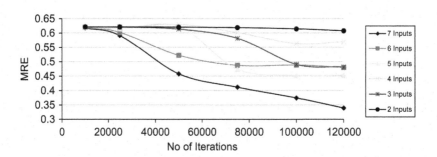

FIGURE 8.23
Variation of MRE with different iterations for various number of inputs.

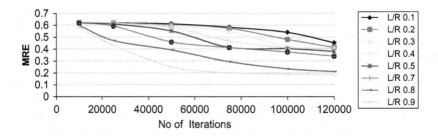

FIGURE 8.24
Variation of MRE with different iterations for various learning rate.

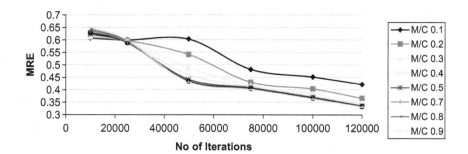

FIGURE 8.25
Variation of MRE with different iterations for various momentum coefficients.

TABLE 8.22

The Input and Output Quantities Used in the Models

Variables	Minimum	Maximum
Input parameters		
20mm Aggregate(Kg/6m3)	3148	3243
12mm Aggregate (Kg/6m3)	3142	3243
Sand (Kg/m3)	4068	4290
Out parameter		
28 day's Compressive strength(Mpa)	29.3	45.8

defined to give numerical meaning to each label. Each membership function identifies the range of input values that corresponds to a label. Typical forms of the membership function with different shapes are presented in Figures 8.26, 8.27 and 8.28.

For each crisp input variable, membership functions are assigned from the following formulas.
   For triangular shape functions,

$$
\begin{aligned}
A(u:\alpha,\beta,\gamma) &= 0 && u < a \\
&= (u-\alpha)/(\beta-\alpha) && \alpha <= u <= \beta \\
&= (\alpha-u)/(\beta-\alpha) && \beta <= u <= \gamma \\
&= 0 && u > \gamma \\
&= 1 && u = \beta
\end{aligned}
$$

For Trapezoidal shape functions,

$$
\begin{aligned}
f(x: a, b, c, d) &= 0 && \text{when } x < a \text{ and } x > d \\
&= (x-a)/(b-a) && \text{when } a <= x <= b \\
&= 1 && \text{when } b <= x <= c \\
&= (d-x)/(d-c) && \text{when } c <= x <= d
\end{aligned}
$$

TABLE 8.23

Fuzzy Rule Formation for Three Fuzzy
Sets for Compressive Strength

Fuzzy Rules	20 mm	12 mm	Sand	Strength
1	L	L	L	L
2	L	L	L	M
3	L	L	M	M
4	L	L	M	H
5	L	L	H	L
6	L	L	H	M
7	L	M	L	M
8	L	M	L	H
9	L	M	M	L
10	L	H	H	H
11	M	L	L	L
12	M	L	L	M
13	M	L	H	L
14	M	M	L	M
15	M	M	H	M
16	H	L	M	H
17	H	L	H	H
18	H	M	H	L
19	H	M	H	M
20	H	H	H	L
21	H	H	H	M

TABLE 8.24

Fuzzy Rule Base for Three Fuzzy Sets for Compressive Strength

Rule Number	Rules
1	**IF** *20mm,12mm* **AND** *Sand* is *LOW* **THEN** *Strength* is *LOW*
4	**IF** *20mm* is *LOW*,12mm is *LOW* **AND** Sand is *MEDIUM* **THEN** *strength is HIGH*
7	**IF** *20mm* is *LOW*,12mm is *MEDIUM* **AND** Sand is *LOW* **THEN** *MEDIUM*
12	**IF** *20mm* is *LOW*, 12mm is *MEDIUM* **AND** sand is *HIGH* **THEN** Strength is *MEDIUM*
30	**IF** *20mm, 12mm* and Sand is *HIGH* **THEN** Strength is *MEDIUM*

For S-shape functions,

$$S(u: \alpha, \beta, \gamma) = 0 \qquad\qquad u < \alpha$$
$$= 2\left[(u-\alpha)/(\gamma-\alpha)\right]^2 \qquad \alpha < u <= \beta$$
$$= 1 - 2\left[(u-\gamma)/(\gamma-\alpha)\right]^2 \qquad \beta < u <= \gamma$$
$$= 1 \qquad\qquad u > \gamma$$

TABLE 8.25

Zone for Three Fuzzy Sets

	Low	Medium	High
Input			
20 mm Aggregate(Kg)	3142–3195.5	3171.75–3219.25	3195.5–3243
12mm Aggregate(Kg)	3142–3192.5	3167.25–3217.75	3192.5–3243
Sand(Kg)	4068–4179	4123.5–4234.5	4179–4290
Output			
Compressive strength(Mpa)	29.3–37.55	33.425–41.675	37.55–45.8

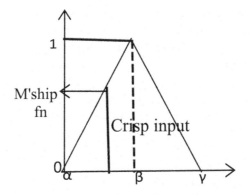

FIGURE 8.26
Typical form of triangular membership function.

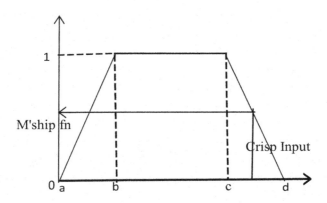

FIGURE 8.27
Typical form of trapezoidal membership function.

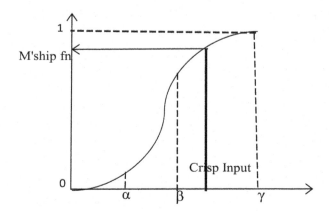

FIGURE 8.28
Typical form of S-shape membership function.

>   **Fuzzy rule inference:** For this analysis, inferencing is done with the Min-Max and Product principles. In min-max inferencing method, the output membership function is clipped off at a height corresponding to the rule premise's computed degree of truth and in product inferencing the output membership function is scaled by the rule premises' computed degree of truth. The diagrammatic representation of fuzzy inference process with min inference method is shown in Figure 8.29.
>
>   **Defuzzification:** This stage is used to convert the fuzzy output set to a crisp number. Two of the more common techniques are the center of gravity and mean of maximum methods.

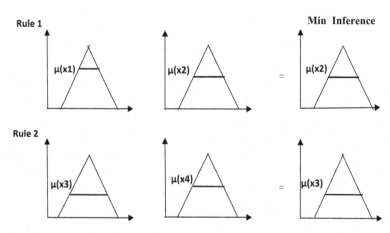

FIGURE 8.29
Fuzzy rule inference process.

In this, defuzzification is done with three different shapes. For triangular and trapezoidal shape, mean of maximum defuzzification method is employed. For computing membership function and for fuzzy inferencing, symmetrical triangular and trapezoidal shapes are considered. Whereas for defuzzification, unsymmetrical triangular and trapezoidal shapes are considered. For s-shape, the singleton method is employed.

In the mean of maxima method, the crisp value of the output variable is the maximum truth-value (membership weight) of the fuzzy subset. Fuzzy singleton is a fuzzy set whose support is a single point in the fuzzy with a membership function of one. Figure 8.30 illustrates the complete process of defuzzification method with mean of maxima.

Fuzzy analysis with specimen calculation: In this analysis, using batch mix ingredients as the input parameters and strength as the output parameter, strength analysis has been carried out. The stepwise procedure followed in this fuzzy analysis work is explained by taking an example with different shape of membership functions. The specimen calculation for the Low-Low-Low-Medium rule for strength analysis is explained for different shape with defuzzification method using the following procedures and analysis for other rules, the same procedure is followed.

Step 1: Fuzzification

Considering one fuzzy set in the rule Low-Low-Low-Medium (20mm-12mm-Sand-Strength) of strength analysis, in the table; the row containing the rule represents the input value quantity of each input variable and corresponding strength of output

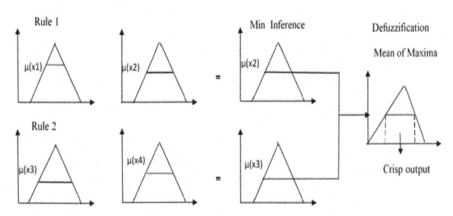

FIGURE 8.30
Defuzzification process.

variable. Membership functions calculated for each variable with different shape is presented in Table 8.26.

Step 2: Fuzzy rule Inferencing

In fuzzy inference, using operator AND, minimum value of the membership function is calculated. Using operator OR, maximum value of membership function is derived. Using operator PRODUCT, multiplication of membership function of the fuzzy set is calculated. Tables 8.27 to 8.29 shows the inferencing procedure for the rule.

Step 3: Defuzzification

In this step of analysis, the fuzzy input is converted into a crisp output. Rule is defuzzified with a corresponding inference value with respect to a corresponding shape and corresponding range value. If the same fuzzy set is falling in another rule and the corresponding minimum value of inference is found out, then among these two rules, then whichever value is minimum is selected and defuzzified with corresponding range value with respect to the corresponding shape. The same procedure is followed for other inference methods and shapes.

TABLE 8.26

Membership Functions for the Rule L-L-L-L with Different Shape

	Input			Output
	Low	Low	Low	Medium
M'ship fns	20 mm	12 mm	Sand	Compressive Strength
Rule	3173	3177	4069	34.70
Triangle	0.9473	0.6138	0.01801	34.70
Trapezoidal	.972	0.91176	0.02702	34.70
S-shape	0.55125	0.81159	0.00016	34.70

TABLE 8.27

Fuzzy Inferencing for Triangle Shape Membership Functions

	Low	Low	Low	Medium
Fuzzy Rule Inference for Triangle	20 mm	12 mm	Sand	Compressive Strength (Mpa)
	3173	3177	4069	34.70
Min	0.9473	0.6138	0.01801	0.01801
Max	0.9473	0.6138	0.01801	0.9473
Product	0.9473	0.6138	0.01801	0.0105

TABLE 8.28

Fuzzy Inferencing for Trapezoidal Shape Membership Functions

Fuzzy Rule Inference for Trapezoidal	Low 20 mm 3173	Low 12 mm 3177	Low Sand 4069	Medium Compressive Strength (MPa) 10419
Min	.972	0.91176	0.02702	0.02702
Max	.972	0.91176	0.02702	.972
Product	.972	0.91176	0.02702	0.0246

TABLE 8.29

Fuzzy Inferencing for S-Shape Membership Functions

Fuzzy Rule Inference for S-shape	Low 20 mm 3173	Low 12 mm 3177	Low Sand 4069	Medium Compressive Strength (MPa) 34.70
Min	0.55125	0.81159	0.00016	0.00016
Max	0.55125	0.81159	0.00016	0.81159
Product	0.55125	0.81159	0.00016	0.0000716

As per the methodology mentioned in Chapter 3, taking an example of one rule Low-Low-Low-Medium, the results and graphs of the rule analyzed with respect to different shapes are mentioned in the various tables. Various defuzzification results are presented in Tables 8.30, 8.31, and 8.32.

In Table 8.32 and Figure 8.31, results of target and computed results for rule L-L-L-M for mean of maxima for min inference with different shape of membership functions are shown. From the table and the graph, it is

TABLE 8.30

Results of Analysis with Mean of Maxima Defuzzification

FIS Type	Defuzzification Method	Fuzzy Inference	M'ship Function	MSE	MRE	RMSE	Cc
Mamdani	Mean of Maxima	Min	Triangle	5.75	5.95	2.38	0.94
			Trapezoid	2.49	3.01	1.62	0.95
			S-shape	9.36	5.68	2.95	0.77
		Max	Triangle	7.97	7.20	2.82	0.95
			Trapezoid	0.27	0.14	0.18	0.972
			S-shape	9.09	15.6	8.95	0.65
		Product	Triangle	5.22	5.84	2.28	0.91
			Trapezoid	2.42	2.68	1.42	0.92
			S-shape	12.94	7.01	3.49	0.78

TABLE 8.31

Results of Analysis with Mean of Minima Defuzzification

FIS Type	Defuzzification Method	Fuzzy Inference	M'ship Function	MSE	MRE	RMSE	Cc
Mamdani	Mean of Minima	Min	Triangle	4.37	5.30	1.97	0.95
			Trapezoid	3.98	4.79	1.83	0.95
			S-shape	15.42	8.44	4.01	0.85
		Max	Triangle	7.91	7.14	2.78	0.95
			Trapezoid	1.30	1.23	1.28	0.957
			S-shape	19.72	8.51	4.23	0.68
		Product	Triangle	4.21	4.95	2.04	0.92
			Trapezoid	3.85	4.65	1.94	0.9
			S-shape	20.10	10.46	4.42	0.89

TABLE 8.32

Defuzzification with Mean of Maxima with Min Inference

Data No.	Target	Triangle	Trapezoid	S-Shape
1	34.7	35.52	35.51	29.41
2	33.5	34.02	34.03	32.95
3	36.9	36.82	36.9	34.54
4	34.2	36.77	34.2	35.03
5	34.6	36.64	34.6	33.98
6	33.7	36.83	33.7	34.76
7	41.22	43.07	42.22	41.19
8	34.22	34.62	34.22	32.42
9	38.22	40.78	38.22	38.00
10	40.8	42.83	40.01	41.49
11	40.9	40.87	40.9	38.38
12	41.6	42.84	43.6	40.60
13	41.7	40.83	42	38.22
14	42.8	42.84	42.8	40.75
15	34.2	34.79	34.2	34.65
16	41.2	42.95	43.2	41.02

observed that results from triangular shape are overestimated, and the results of other shapes following the path of target value but slightly varying in the values.

The statistical parameters of the predicted and actual values of compressive strength are presented in Table 8.33. It is observed that trapezoidal shape max and mean of maxima were employed as inference operator and defuzzification methods, respectively, giving good results with correlation coefficient (Cc) as 0.97.

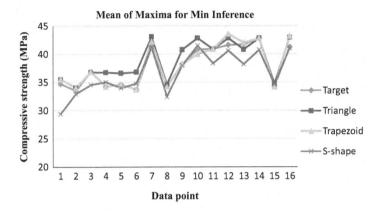

FIGURE 8.31
Comparison of target strength with computed fuzzy model results for rule L-L-L-M for mean of maxima for min inference.

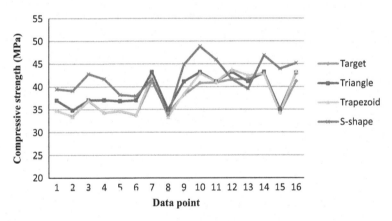

FIGURE 8.32
Comparison of target strength with computed fuzzy model result for rule L-L -L-M for mean of maxima for max inference.

8.5 In the Field of Coastal and Marine Engineering

A significant wave height is defined as the average height of the one-third highest waves and it is about equal to the average height of the waves as estimated by an experienced observer.

8.5.1 Need of Forecasting

Hs is a significant input parameter for the design of harbors, coastal structures, offshore structures, defense purposes, planning, operations, coastal

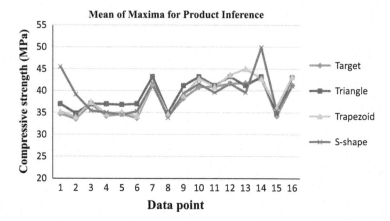

FIGURE 8.33
Comparison of target strength with computed fuzzy model results for rule L-L-L-M for mean of maxima for product inference.

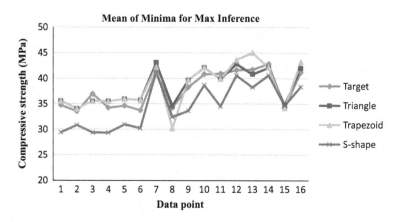

FIGURE 8.34
Comparison of target strength with computed fuzzy mode result for rule L-L-L-M for mean of minima for max inference.

erosion, modeling of sediment transport, oil spill modeling, wave energy estimation, etc.

8.5.2 Results from ANN Model

To forecast significant wave height various input scenarios [(1) Hs(t), (2) Hs(t), Hs(t-1), (3) Hs(t), Hs(t-1), Hs(t-2), and (4) Hs(t), Hs(t-1), Hs(t-2), Hs(t-3)] of previous time steps wave height up to 12 hours are considered in this model (ANN). The results obtained from different input scenarios are presented in Table 8.34.

It is clear from Table 8.34 that the input scenario-4 is producing good results compare to other input scenarios, but this improvement was observed to be

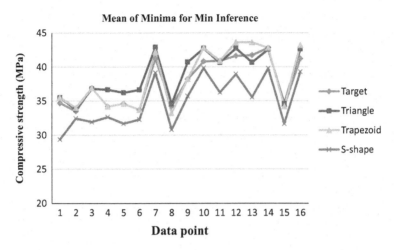

FIGURE 8.35
Comparison of target strength with computed fuzzy model result for rule L-L -L-M for mean of minima for min inference.

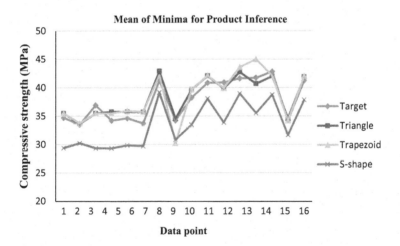

FIGURE 8.36
Comparison of target strength with computed fuzzy model result for rule L-L -L-M for mean of minima for product inference.

very minimal. The results obtained from scenario-4 are presented in the form of time series plots and scatters plots for various lead times.

The time series trend plot and scatter plot for 3h and 6h predictions are presented in Figures 8.37 and 8.38. The predicted values from the ANN almost match the observed values in 3 h, but some differences are observed in 6 h predictions of the previous plots.

But in the 12th h and 24th h predictions, the ANN performance started to decrease as seen in Figures 8.39 and 8.40. The upper values more than 4 m were not covered in the 24 h predictions. Figure 8.41 shows the 48-hour predictions.

TABLE 8.33

Predicted and Actual Compressive Strengths

S. No.	Input Parameters			Output Parameter	Fuzzy Results		
	20 mm	12 mm	Sand	Target Strength	Triangle	Trapezoid	S-Shape
1	3173	3177	4068	34.7	36.98	34.77	39.5
2	3169	3183	4144	33.5	34.82	33.4	39.1
3	3172	3167	4115	36.9	37.05	37	42.8
4	3170	3172	4124	34.2	37.05	34.3	41.7
5	3168	3174	4110	34.6	36.81	34.7	38.2
6	3171	3171	4117	33.7	37.01	33.79	37.9
7	3169	3167	4117	41.2	43.23	42.32	41.6
8	3171	3167	4110	34.2	34.98	33.32	33.8
9	3171	3168	4109	38.2	41.14	38.42	44.8
10	3171	3174	4121	40.8	43.2	45.9	48.8
11	3170	3167	4109	40.9	41.17	41.13	45.9
12	3170	3167	4109	41.6	43.23	43.7	41.6
13	3169	3167	4111	41.7	41.17	45.44	39.6
14	3172	3186	4111	42.8	43.23	42.9	46.9
15	3169	3167	4116	34.2	34.98	34.3	43.9
16	3168	3172	4116	41.2	43.04	43.3	45.1
17	3167	3167	4133	43.5	43.07	43.37	42.77
18	3168	3170	4192	44.3	43.17	44.43	41.88
19	3167	3172	4193	38.2	37.05	37.38	46.7
20	3172	3186	4111	40	41.17	40.88	42.09

8.6 In the Field of Environmental Engineering

8.6.1 Dew Point Temperature Modeling

The dew point temperature is the temperature at which the moisture in the air begins to condense into dew or water droplets. The accurate estimation of the dew point temperature is very important as it controls the heat stress on humans and detects fluctuations in evaporation rates and humidity trends. The dew point temperature is a significant parameter required in various hydrological, climatological, and agronomical related research. This study proposes SVM and ELM models for the estimation of daily dew point temperature. The daily measured weather data (wet-bulb temperature, relative humidity, vapor pressure, and dew point temperature) of humid and semi-arid regions of India were used for model development. The statistical indices, namely mean absolute

TABLE 8.34

Test Results of ANN Model for Different Input Scenarios

		Hs (t)	Hs(t-1), Hs(t)	Hs(t-2), Hs(t-1), Hs(t)	Hs(t-3), Hs(t-2), Hs(t-1), Hs(t)
	MAE (m)	0.316	0.317	0.313	0.308
	MRE (%)	8.969	8.965	8.855	8.698
3h	RMSE (m)	0.435	0.437	0.431	0.423
	R^2	0.901	0.901	0.903	0.907
	MAE (m)	0.469	0.472	0.466	0.459
6h	MRE (%)	13.343	13.463	13.269	13.013
	RMSE (m)	0.650	0.650	0.640	0.631
	R^2	0.780	0.781	0.787	0.794
	MAE (m)	0.689	0.693	0.691	0.673
12h	MRE (%)	19.522	19.749	19.646	19.014
	RMSE (m)	0.955	0.955	0.950	0.934
	R^2	0.526	0.526	0.532	0.550
	MAE (m)	0.974	0.978	0.969	0.970
24h	MRE (%)	28.125	28.883	27.813	27.836
	RMSE (m)	1.304	1.300	1.298	1.296
	R^2	0.151	0.153	0.165	0.167
	MAE (m)	1.133	1.134	1.128	1.135
48h	MRE (%)	32.782	32.888	32.982	33.182
	RMSE (m)	1.469	1.469	1.466	1.471
	R^2	0.012	0.014	0.014	0.013

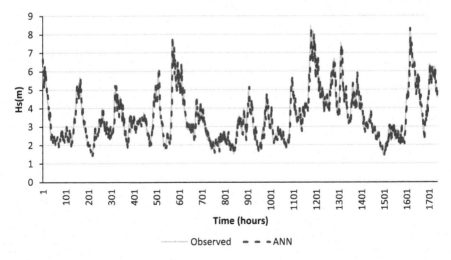

FIGURE 8.37
Model performance during testing for third hour prediction.

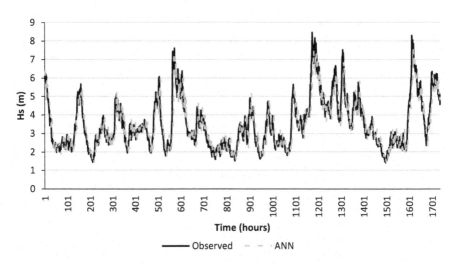

FIGURE 8.38
Model performance during testing for sixth hour prediction.

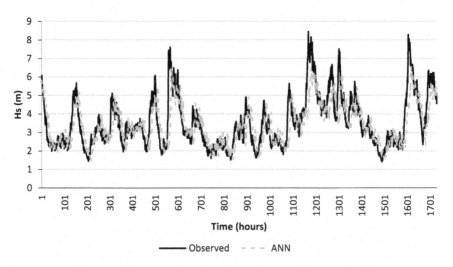

FIGURE 8.39
Model performance during testing for twelfth hour prediction.

error (MAE), root mean square error (RMSE), Nash–Sutcliffe efficiency (NSE) were adopted to evaluate the performances of these two models. The merit of the extreme learning machine (ELM) model is evaluated against support vector machine (SVM) technique in the estimation of dew point temperature. The proposed ELM models demonstrated much greater capability than the SVM models in the estimation of daily dew point temperature.

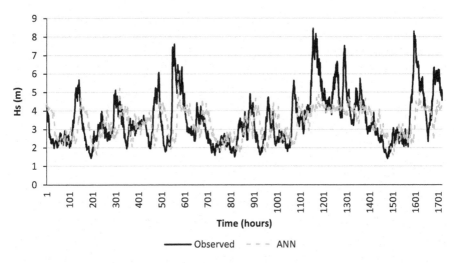

FIGURE 8.40
Model performance during testing for twenty-fourth hour prediction.

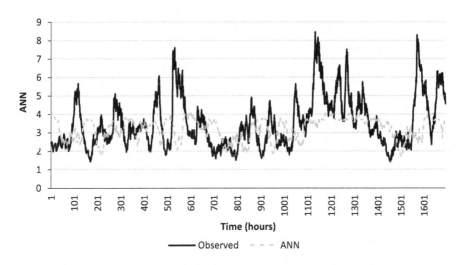

FIGURE 8.41
Model performance during testing for forty-eighth hour prediction.

8.6.2 Air Temperature Modeling Using Air Pollution and Meteorological Parameters

Ambient temperature is an important variable for energy balance and water cycle of the soil/vegetation/atmosphere continuum. Owing to that, the determination of one of the meteorological parameters like ambient temperature is required for various fields like climate monitoring, drought detection, and the environment. Since average air temperature is one of the most fundamental elements of climate variation in an area, intense work has been earmarked to the investigation of the average air temperature behavior in different times and location scales. Predictions of temperature are of special importance in the fields of water resources planning and management, irrigation networks, tourism, and everyday life issues.

Due to the dynamic nature of the atmosphere, it is difficult to predict ambient air temperature data accurately. Various techniques, like linear regression, auto regression, multi-layer perceptron, radial basis function networks, are applied to predict atmospheric parameters like temperature, wind speed, rainfall, meteorological pollution, etc.

8.6.2.1 Performance Analysis of Models for Seven Stations (Meteorological Parameters Only) ANFIS Model

The monthly average observed meteorological data of seven stations (132 months) were used in this study. Among those 132 months of data of each parameter, 99 months of data were used for training and 33 months of data were used for testing. Here, the original raw data of parameters like rainfall, wind speed, humidity, and sunshine hours are used for the input of ANFIS model (Sugeno first order with 16 fuzzy rules and Gbell membership function) for both training and testing. Average air temperature data were used as output for both training and testing. Results are shown in Table 8.38. In case of ANFIS model testing, CC is less than 0.5 and RMSE values are more than 6.6, which reveals its poor performance. In SI for testing, values are more than 0.2, which is beyond the acceptable limit in terms of accuracy. When ANFIS model results are compared among the stations, Hiriyur stations having better results in terms of CC, RMSE, and SI as shown in Table 8.38. For ANFIS, station Shimoga and Honnali having less CC and high RMSE value, this may be due to the higher degree of nonlinearity and presence of noisy data.

A time-series plots for all the seven stations are shown in Figures 8.42, 8.43, 8.44, 8.45, 8.46, 8.47, and 8.48. Here too, good agreement between estimated and measured air temperature are observed throughout the testing period for ANFIS model. Also, during the summer months (April to May), a moderate error between observed and estimated is identified. This can be termed as highly overestimated. For a lower temperature region, estimated temperature was found to be somewhat nearer to measured air temperature.

8.6.2.2 SVM model

Statistical performance indices computed using the modeled and observed values of testing data for the SVM models are presented in Table 8.39. The performance of SVM depends on the reliable setting of SVM and kernel parameters. In developing SVM models, initially parameters are randomly selected by coarse grain search (i.e., for C = 100, 200, 300, …, 2000; ε = 0.5, 1, …, 2; and

TABLE 8.35

Optimal Values of RBF Kernel-Based SVM Hyperparameters

SVM Models		C	γ	ε	No. of Support Vectors
Bajpe Station	3rd Hour Model	22	8	0.01	235
	12th Hour Model	28	7	0.01	254
Hyderabad Station	3rd Hour Model	37	12	0.01	262
	12th Hour Model	41	14	0.01	269

TABLE 8.36

Testing Performance of SVM and ELM Models at Bajpe Station

Model	Model Parameters	RMSE (°C)	MAE (°C)	NSE
3rd hr. SVM	28, 8, 0.01	0.48	0.21	0.52
3rd hr. ELM	3-40-1	0.38	0.04	0.69
12th hr. SVM	28, 7, 0.01	0.52	0.28	0.62
12th hr. ELM	3-90-1	0.10	0.02	0.90

TABLE 8.37

Testing Performance of SVM and ELM Models at Hyderabad Station

Model	Model Parameters	RMSE (°C)	MAE (°C)	NSE
3rd hr. SVM	28, 8, 0.01	2.36	1.04	0.63
3rd hr. ELM	3-50-1	0.63	0.32	0.95
12th hr. SVM	28, 7, 0.01	1.98	1.05	0.82
12th hr. ELM	3-70-1	0.59	0.14	0.97

TABLE 8.38

ANFIS Model Performance

Sl. No.	Stations	ANFIS Model		
		CC	RMSE (°C)	SI
1	SHIMOGA	0.097	20.980	0.810
2	HONNALI	0.438	85.200	3.310
3	B.R.PROJECT	0.280	6.640	0.260
4	DAVANGERE	0.520	10.160	0.380
5	LINGANAMAKKI	0.160	31.890	1.200
6	HIRIYUR	**0.409**	**6.700**	**0.270**
7	RAIPURA	0.409	6.705	0.270

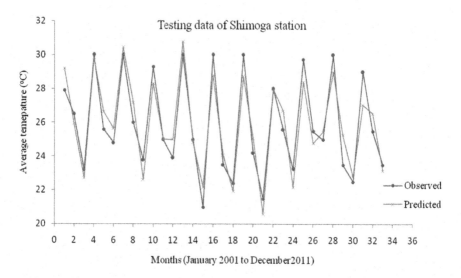

FIGURE 8.42
ANFIS Model performance at Shimoga station.

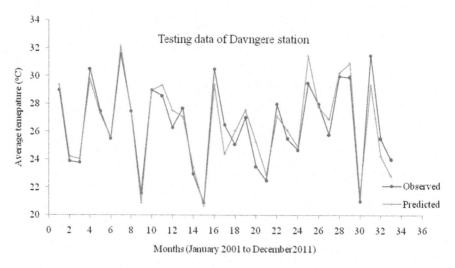

FIGURE 8.43
ANFIS Model performance at Davngere station.

$d = 1, 2, \ldots, 6$) to identify the near optimal values, and then a fine grain search (i.e., for $C = 50, 100, 500, \ldots, 5000$; $\varepsilon = 0.000001, \ldots, 2$; and $d = 1, 2, \ldots, 6$) is done to identify the final optimal values. The final optimum values (i.e., for nsv = 99; $C = 50$; $\varepsilon = 0.1$; and $d = 0.5$) of SVM and polynomial as kernel function. In the case of the SVM model for seven meteorological stations, CC value was found less than 0.5, with a RMSE more than 2.310 for testing, and SI values are less than 0.1, which shows inferior performance.

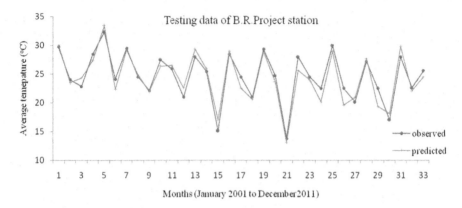

FIGURE 8.44
ANFIS Model performance at BR Project station.

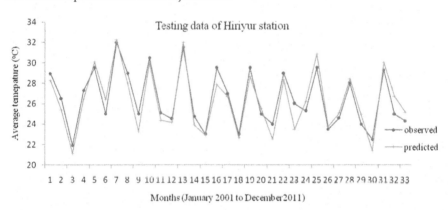

FIGURE 8.45
ANFIS Model performance at Hiriyur station.

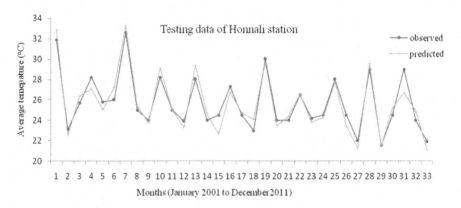

FIGURE 8.46
ANFIS Model performance at Honnali station.

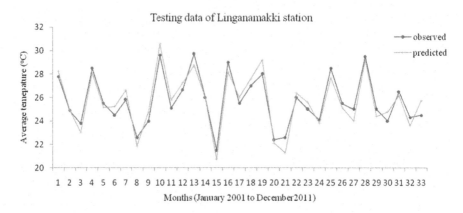

FIGURE 8.47
ANFIS Model performance at Linganamakki station.

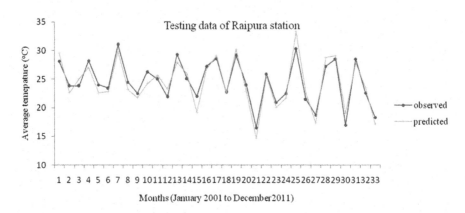

FIGURE 8.48
ANFIS Model performance at Raipura station.

TABLE 8.39

SVM Model Performance

Sl. No.	Stations	SVM Model		
		CC	RMSE (°C)	SI
1	**SHIMOGA**	0.180	2.720	0.100
2	**HONNALI**	**0.530**	**2.310**	**0.080**
3	**B.R.PROJECT**	0.010	4.470	0.180
4	**DAVANGERE**	−0.200	3.290	0.100
5	**LINGANAMAKKI**	-0.030	3.010	0.110
6	**HIRIYUR**	-0.070	3.810	0.140
7	**RAIPURA**	0.310	3.630	0.140

Also, during summer months (April to May), a huge deviation between observed and estimated is identified, which can be termed as highly overestimated. For a lower temperature region, estimated temperature are found to be somewhat nearer to measured air temperature.

8.7 In the Field of Structural Engineering

Many researchers have developed neural dynamics models for the optimum design of structures. It's been shown how neural networks can be used for estimating resource requirements at the conceptual stage of design. Neural networks are versatile tools for analogy-based solutions. They use ANN to predict the ultimate shear strength of reinforced-concrete deep beams. The result of the ANN was compared with various empirical relationships and proved that ANNs provide a reliable prediction of shear strength. Many researchers have also developed neural networks for the optimum design of simply supported concrete beams and reinforced fibre Porterhouse concrete beams, with proven superiority of ANNs compared with conventional design techniques. Also, it is explored the applications of ANN to predict the compressive strength and corresponding strain of circular concrete columns. And the ANN model has been developed to predict the shear strength of reinforced-concrete beams with and without web reinforcement as previously mentioned.

8.8 In the Field of Transportation Engineering

Soft computing (SC) has played a major role in the success of Intelligent Traffic System (ITS) in recent years, especially because of the larger amount of data provided and collected from several sources by the different stakeholders involved in these systems, such as governments, industry, and citizens. ITS is an environment very appropriately applied to SC techniques because it handles most of the features for which SC was designed for. For example, the sensors used usually present imprecision in their measures; traffic is strongly affected by factors that pose high uncertainty, such as weather; and the decision-making should take into account drivers' or users' preferences subject to vagueness and subjectivity. For this reason, techniques like fuzzy sets, neural networks, metaheuristics, and probabilistic reasoning have been widely used by the research community in ITS.

8.8.1 Soft Computing for Traffic Congestion Prediction

In today's societies, transportation is a fundamental condition in everyone's life. The preferred mode of transport in a typical day is the car, well above urban public transport. This issue, added to the fact that the number of vehicles per capita has increased in the last ten years, has raised the efficient of transportation to the level of a fundamental condition, especially in big cities. Due to that, investment in transportation research on improvement of current transportation systems has led to projects related to this theme. As mentioned in Lopez-Garcia et al. (2016), the economic costs of traffic congestion will increase by about 50% by 2050, accessibility gap between central and peripheral areas will widen, and the social costs of accidents and pollution will continue to rise.

For all these reasons, road trips are a key point of ITS, due to the importance in daily life not only for people but also for transportation companies. Inside this subject, one field where different techniques are being used during the last few years with a high impact and reliable performance is the forecasting of the traffic state in freeway and urban scenarios. One of the principal challenges in this field is to predict, with a certain level of confidence, possible traffic jams on a short-term horizon.

The principal advantages of the successful prediction of traffic jams is the adaptation of decision-making in the exact moment when different events that may affect traffic, like accidents, occur. Another advantage is the capability of calculating not only the travel time but also of planning the route to follow before its beginning.

If the user knows the probability of finding a traffic jam in its route, he/she can avoid it by changing the route before or even during the journey. In a general way, the successful prediction of traffic jams can lead to the decrease of travel time, the reduction of CO_2 emissions, as well as fuel consumption, or the decrease of acoustic contamination in urban and freeway environments.

8.8.2 Neural Networks in Traffic Congestion Prediction

In the last few years, traffic congestion prediction is one of the fields where NNs has increased, as seen in the literature. For example, an NN was applied to predict traffic congestion using past traffic data. Volume, speed, density, and both time and day of the week were used as input variables. The model was validated using rural highway traffic. Another case was presented wherein a Gaussian mixture model clustering was combined with an NN to create an urban traffic flow prediction system. The system forecasts traffic flow by combining road geographical and environmental factors with traffic flow properties obtained by the use of detectors. Another type of NN, called Back Propagation NN (BPNN) was used to forecast campus traffic congestion levels. The results were compared with a Markov model, where BPNN achieved higher accuracy and more stable performance. Deep belief networks were used to enhance prediction accuracy using weather conditions.

The study had two objectives: to investigate a correlation between weather parameters and traffic flow and to improve traffic flow prediction accuracy.

8.8.3 Fuzzy Systems in Traffic Congestion Forecasting

Fuzzy logic allows for the processing of inaccurate information using IF–THEN rules, which helps in the interpretation of the final model. One of the most used and well-known types of fuzzy systems are fuzzy rule-based systems (FRBS), which can be divided into Mamdani and Takagi–Sugeno–Kang (TSK) systems. Another class of systems, based on the previous ones, are called Hierarchical FRBS (HFRBS). This class of systems is comprised of several FRBSs which are joined in a way that the output of one of them is the input of other. Depending of the structure of the hierarchy, those systems can be divided into parallel, serial, and hybrid.

In traffic congestion prediction, HFRBSs have been used to develop a congestion prediction system employing a large number of input variables. In this chapter, a steady-state GA is applied to tune the different parts of the FRBSs. Zheng et al. (2014), used a hybrid algorithm that combines GA and the cross-entropy method to tune an HFRBS in order to predict congestion on a freeway in California with time horizons of 5, 15, and 30 min. An extension of that work is presented in Yu et al. (2016), wherein state-of-the-art techniques are compared with the results obtained by the tuned HFRBSs in different traffic congestion data sets. (Yu et al,2016)

8.8.4 Soft Computing in Vehicle Routing Problems

Another field in which soft computing techniques have demonstrated outstanding performance is related to vehicle routing problems. Nowadays, route planning is a widely studied field in which the most used and well-known problems are the traveling salesman problem (TSP), and the vehicle routing problem (VRP), the focus of many studies. The reasons for the importance and popularity of studying these problems are two-fold: the scientific aspect and the social one. On the one hand, being NP-Hard, most of the problems arising in this field have great complexity. For this reason, their resolution poses a major challenge for the scientific community. On the other hand, routing problems are usually built to address a real-world situation related to logistics or transportation. This is the main reason because its resolution entails a profit—either a business or social one.

Furthermore, several approaches can be found in the scientific community to tackle this kind of problem in an efficient way. In this sense, two of the most successful method are the exact methods—heuristics and metaheuristics.

8.9 Other Applications

8.9.1 Soil Hydraulic Conductivity Modeling

Soft computing techniques, namely ANFIS, SVM, and ELM were used to develop pedotransfer function for the data measured at various depths below ground level and three different locations. All of the data set from three sites and three depths (900) were split into smaller data sets for the modeling purpose into six subsets, which are described below. Each sub-data set has 300 data for all the parameters.

1. College station at all three depths (15cm, 30cm, and 45cm)
2. Mulegaon station at all three depths (15cm, 30cm, and 45cm)
3. Punanaka station at all three depths (15cm, 30cm, and 45cm)
4. 15cm depth at all three locations (College, Mulegaon, and Punanaka)
5. 30cm depth at all three locations (College, Mulegaon, and Punanaka)
6. 45cm depth at all three locations (College, Mulegaon, and Punanaka)

Statistical analyzes were carried out for all the samples. Results of statistical analysis were discussed; for modeling purposes, data need to be normally distributed, which was checked by using QQ plot. It was shown that the Kfs (field saturated hydraulic conductivity) is log normalized, thus all data were log transformed). Data-driven techniques usually perform well when the range of values of all parameters are the same; to meet this requirement further data were normalized by using an equation as discussed.

Normalized data at each sampling (location/depth) were split into two data sets; one was used for training the network and validity of the model after the training was tested using the other subset. Each data set (300 samples) were divided into six combinations for training and validation of each of these models. Optimum modeling parameters of each model were determined for all these six combinations, and the best model among them was determined based on the performance criteria. The details are discussed below.

- **Training/testing data set:** For each of three models the following trials were carried out by segregating data into training and testing (validation) to determine the ideal model. For segregating data, they were arranged in order (descending/ascending), and then every nth element (3rd, 6th, 9th, etc.) resulting from this arrangement was segregated to produce a data set for validation.

Development of Models

- **ELM model:** Three-layer architecture was adopted for the ELM model development. The first layer (input) used various soil

TABLE 8.40

Sample Distribution for Training and Validating Model

Trial	Training Data Set		Testing Data Set	
	% of Total Data	No. of Data	% of Total Data	No. of Data
I	90	270	10	30
II	85	255	15	45
III	80	240	20	60
IV	75	225	25	75
V	70	210	30	90
VI	67	200	33	100

parameters as inputs. The output layer had one neuron representing the estimated saturated hydraulic conductivity (Kfs). For the hidden layer a maximum of 100 neurons were tested for each model. Initially, one neuron was selected, and subsequently the number of neurons was gradually increased up to 100 by an interval of one. Radial basis activation function was employed for all ELM models tested.

- **SVM model**: In this study, the SVM regression was performed in two stages, training and testing. Data were normalized between 0.05 and 0.95 before modeling. The normalized data set was used to develop the SVM regression model. The output results obtained were denormalized. During the training stage, SVM parameters C (cost function), kernel width (γ), and insensitive value (ε) are optimized by performing a thorough grid search. These hyperparameters are interdependent and, thus, the possible combination of these three parameters will be chosen based on the grid search method.

- **Grid search method**: A time-consuming, two-step grid search method, was used wherein an initially coarse grid search was applied maintaining a wide range for the parameters with large

TABLE 8.41

ELM Model Parameters

Model	Hidden Layer	Number of Neurons in the Hidden Layer	Data Used for Training
ELM-CO	01	15	90 %
ELM-MU	01	28	70 %
ELM-PN	01	10	67 %
ELM-15	01	12	75 %
ELM-30	01	11	75 %
ELM-45	01	15	80 %

increments (say 2^{-15} to 2^{15} with increments of two in the exponent) to obtain the best region of these parameters. Then, in that region, a finer grid search was performed for each parameter. The hyper-parameters were optimized by estimating the mean square error for every possible combination of these three parameters; the combination of hyperparameters which results in the minimum value of mean square error during training will be regarded as optimum hyperparameters. To avoid the danger of overfitting, the four-fold cross-validation approach was used during the training phase. LIBSVM software is used for analysis and calculation.

- **ANFIS model:** In this model, the fuzzy c-means (FCM) clustering algorithm is used to divide the data set into clusters. FCM is a soft data clustering technique wherein each data point belongs to a cluster that is, to some degree, specified by a membership grade (0 to 1). The algorithm used for clustering is described:

1. $U=[u_{ij}]$ matrix, $U^{(0)}$

   At k – step: calculate the centers vectors $C^{(k)}=[c_j]$ with $U^{(k)}$

$$C_i = \frac{\sum_{j=1}^{n} u_{ij}^m x_j}{\sum_{j=1}^{n} u_{ij}^m}$$

2. Update $U^{(k)}$, $U^{(k-1)}$

$$d_{ij} = \sqrt{\sum_{i=1}^{n} (x_i - c_i)}$$

$$u_{ij} = \frac{1}{\sum_{k=1}^{c} \left(\frac{d_{ij}}{d_{ki}}\right)^{2/(m-1)}}$$

TABLE 8.42

SVM Model Parameters

Model	Hyper Parameter			Number of Support Vectors	Data Used for Training the Model
	C	Gamma	Epsilon		
SVM-CO	34.67	0.0323	0.00097656	38	90 %
SVM-MU	18.34	0.0418	0.0000488	35	70 %
SVM-PN	0.5548	16.2234	0.0078	33	67 %
SVM-15	0.1250	8.5742	0.0018	28	75 %
SVM-30	10.76	3.427	0.035	34	75 %
SVM-45	12.37	4.867	0.00043	42	80%

3. If $\|U(k+1) - U(k)\| < \varepsilon$ then stop; otherwise return to step 2

where m, any real number greater than 1 (~2) u_{ij} degree of membership of x_i in cluster j x_i is the ith of d (dimensional measured data)

c_j is the d dimension canter of cluster

ε is a termination criterion between 0 and 1,

k are the iteration steps.

This procedure converges to a local minimum or a saddle point of Jm.

In fuzzy c-means algorithm, the number of clusters has to be set arbitrarily; i.e., the number of clusters to be created by the clustering algorithm must be set manually on each algorithm execution; this is done repeatedly until we get an optimal number of clusters based on the objective function. Numbers of clusters were tried manually between 2 and 20 with an increment of one for each data set.

- **Performance evaluation of models:** Performances of the models were tested by using coefficient of correlation (R), mean relative error (MRE), normalized root mean square error [(NRMSE), close to zero is better], and Nash–Sutcliffe efficiency (NSE).

   The performance metrics of the training and testing data sets are shown in Table 8.44 for data set segregated for modeling purposes. Scatter plot between observed and predicted value of the saturated hydraulic conductivity by using ELM, SVM, and ANFIS are presented in Figures 8.49 and 8.50.

- **College model:** R-value for training and testing for ELM model (0.97 and 0.96) performance was found good in comparison to the other two models. MRE of ELM in training is 0.02 (~0) and is less than that for the other two models (Table 8.44). In terms of NSE, ELM performance is also very good during training and testing. NRMSE of SVM and ANFIS are relatively good for training however, for testing

TABLE 8.43

ANFIS Model Parameters

Model	ANFIS Model Parameters		Data Used for Training
	No. of Rules	No. of Clusters	
ANFIS – CO	14	12	90 %
ANFIS – MU	15	18	70 %
ANFIS – PN	13	18	67 %
ANFIS – 15	10	4	75 %
ANFIS – 30	17	6	75 %
ANFIS – 45	15	16	80 %

TABLE 8.44

Performance Comparison of Models for College (90% training, 10% testing)

Model	Training			Testing		
	ELM	SVM	ANFIS	ELM	SVM	ANFIS
R	0.97	0.77	0.78	0.96	0.52	0.58
MRE	0.02	0.07	0.06	0.18	0.60	0.58
RMSE (myr^{-1})	73.01	257.12	217.52	219.03	771.37	652.55
NRMSE	0.02	0.07	0.06	0.06	0.21	0.18
NSE	0.95	0.16	0.60	0.90	0.18	0.15

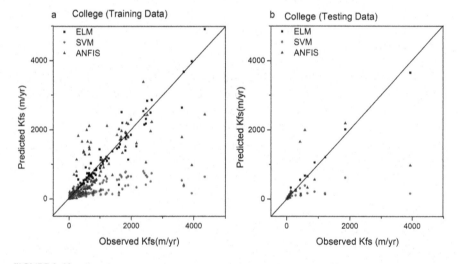

FIGURE 8.49
Scatter plot of predicted Kfs V/S observed Kfs for college station during training (a) and testing (b).

it is more in line with ELM. The scatterplot of observed vs predicted Kfs is shown in Figure 8.48 along with 1:1 line. The predicted values of Kfs by ELM methods are close to the 1:1 line in compared with the other two methods both during training and testing phases. Prediction by SVM method during training and testing were below the 1:1 line indicating it fails to predict the Kfs; the model is underfitting the values both during training and testing phases.

ANFIS prediction are scattered around the 1:1 line; minority values are over-predicted (for smaller values of Kfs) and majority values are under-predicted (for higher values of Kfs). Box plot of observed Kfs and estimated Kfs by the ELM, SVM, and ANFIS methods are shown in Figure 8.49. The median of observed Kfs and estimated Kfs by all methods is roughly the same.

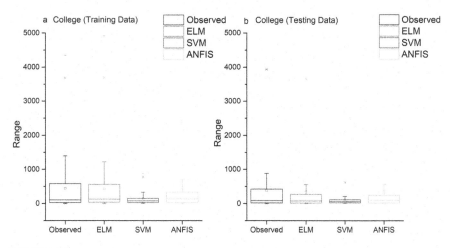

FIGURE 8.50
Box plot of Kfs, for observed and predicted values by ELM, SVM and ANFIS a) during training and b) during testing at college site.

Data distribution in lower quartile (0 – 25%) is dense in observed and predicted Kfs. Box plot of SVM both during training and testing is very compact implying values are very close (dense); this is very effective in predicting lower values but fails to predict higher values of Kfs.

8.9.2 Modeling Pan Evaporation

In the first stage, the SMO-based SVR model is employed for the modeling of pan evaporation using various input combinations. The input parameters identified in the statistical analysis of daily time-series data recorded at both stations selected for the study were used. Although gamma test recognizes the better input–output pattern, justification requires a robust and reliable technique, such as SVR, to be tested with. The model is trained with available daily time series data for five years (2000–04) and tested with two years (2005–07) of data recorded at Bajpe station and daily time series data of six years (1975–81) and tested with three years (1982–84) of data recorded at Bangalore station. As discussed in the previous chapter, the key factor in the success of SVR lies with kernel functions and optimum parameters to fine tune the kernel functions. However, identification of appropriate kernel function in advance of model building is not possible. Therefore, trial and error iterations are essential to judge the suitability of kernel function for the data set. The accuracy of kernels based relies on the selection of the model parameters. The best fitting of models depends upon the number of support vectors generated during model building. The generated number of support vectors should not be more than 70% to avoid overfitting or below 30% to avoid underfitting (Fielding and Bell, 1997). In this study, three types of kernel functions are utilized and their optimal parameters were derived using sequential mean optimization–support vector regression (SMO–SVR) with grid search and CV parameter optimization, as discussed earlier. The optimum results computed with SVR are displayed in Tables 8.45 and 8.46.

8.9.2.1 Performance Evaluation

As disused in the previous chapter, the estimated pan values by SVR models and DWT-SVR models were compared with observed values using four types of performance measures: RMSE, MAE, CC, and NSE. The model performance is said to be optimum with lowest RMSE and MAE, whereas CC and NSE are supposed to be closer to the value one.

Different combinations of input variables that influence the pan evaporation the most were tried on training and testing data sets of both stations. It is clear from the following tables that as the number of influential attributes combines together the model superiority increases. The estimation efficiency is enhanced. In Tables 8.45 and 8.46 corresponding to training and testing data from Bajpe station, there are five types of input combinations are evaluated employing the three types of kernel functions listed in the previous chapter. Initially with a lone attribute of temperature the model results are very poor, showing higher RMSE and MAE and lower values of CC and NAE. But as more influential attributes are added to the model, it shows superior performance. The tables also reveal that the kernel functions played their roles to make the model superior and robust. For the all five input combination scenarios, RBF kernel function showed slightly better performance in comparison to the other two kernel functions in both training and testing period modes. It was also found that Pearson VII universal kernel (PUK) showed similar results to the RBF kernel function. Considering the results phase-wise, it is observed that testing phase results are superior to training results for the estimations computed for Bajpe station.

The modeled results for the Bangalore station training and testing phases are displayed in Table 8.47 and 8.48. Once again, the combination of all influential parameters affected the pan evaporation the most. Kernel estimated results varied with the different combinations and also with training and testing models. The drawback experienced in this set of model performances is the sudden decline in the performance from training to testing data phase. Focusing on the RBF kernel results, it was found that the values of RMSE, MAE, CC, and NAE were 1.009, 0.728, 0.836, and 0.976, respectively, in the training phase; the values of RMSE, MAE, CC, and NAE at testing phase were 1.382, 0.887, 0.681, and 0.959, respectively. There is a rise in values of RMSE and MAE, and the reduced values of CC and NAE confirms such a decline. The reasons behind such decline may be due to the variations in the ranges of variables. It was observed in the statistical analysis of both stations' data that some individual attributes, such as wind speed, rainfall, and relative humidity showed the most variations in their ranges. This may be due to

TABLE 8.45

Statistical Indices of SMO–SVR Models with Combinations of Input Parameters for Bajpe Station (Training Data)

Input Combination	SMO–SVR Polynomial Kernel				SMO–SVR RBF Kernel				SMO–SVR PUK Kernel			
	RMSE (mm)	MAE (mm)	CC	NSE	RMSE (mm)	MAE (mm)	CC	NSE	RMSE (mm)	MAE (mm)	CC	NSE
T	1.321	0.898	0.652	0.931	1.121	0.912	0.747	0.946	1.142	0.926	0.731	0.940
T + W	1.264	0.778	0.720	0.942	0.995	0.878	0.795	0.959	1.002	0.897	0.776	0.956
T + W+ P	1.147	0.782	0.745	0.953	0.982	0.722	0.815	0.965	0.991	0.825	0.794	0.967
T + W + P + Rh	1.132	0.768	0.781	0.961	0.961	0.692	0.827	0.973	0.963	0.695	0.821	0.970
T + W + P + Rh + Sh	1.032	0.759	0.754	0.967	**0.941**	**0.687**	**0.832**	**0.977**	0.906	0.650	0.845	0.981

TABLE 8.46

Statistical Indices of SMO–SVR Models with Combinations of Input Parameters for Bajpe Station (Testing Data)

Input Combination	SMO–SVR Polynomial Kernel			SMO–SVR RBF Kernel			SMO–SVR PUK Kernel		
	RMSE (mm)	MAE (mm)	CC	NSE	RMSE (mm)	MAE (mm)	CC	NSE	RMSE (mm)	MAE (mm)	CC	NSE
T	1.348	1.145	0.611	0.925	1.168	0.929	0.757	0.955	1.142	0.951	0.721	0.951
T + W	1.214	0.926	0.655	0.940	1.112	0.878	0.735	0.959	1.002	0.897	0.731	0.956
T + W + P	1.151	0.898	0.754	0.951	1.024	0.915	0.785	0.962	1.121	0.922	0.781	0.967
T + W + P + Rh	1.122	0.819	0.785	0.969	0.997	0.793	0.817	0.977	1.005	0.802	0.821	0.973
T + W + P + Rh + Sh	0.997	0.770	0.834	0.972	**0.990**	**0.763**	**0.839**	**0.985**	0.981	0.761	0.838	0.977

TABLE 8.47

Statistical Indices of SMO–SVR Models with Combinations of Input Parameters for Bangalore Station (Training Data)

Input Combination	SMO–SVR Polynomial Kernel				SMO–SVR RBF Kernel				SMO–SVR PUK Kernel			
	RMSE (mm)	MAE (mm)	CC	NSE	RMSE (mm)	MAE (mm)	CC	NSE	RMSE (mm)	MAE (mm)	CC	NSE
T	1.485	1.167	0.592	0.948	1.427	1.089	0.632	0.952	1.432	1.096	0.629	0.952
T + W	1.463	1.147	0.610	0.949	1.327	1.041	0.667	0.955	1.400	1.069	0.651	0.954
T + W+ P	1.413	1.088	0.644	0.953	1.249	0.929	0.736	0.963	1.344	1.013	0.686	0.957
T + W + P + Rh	1.108	0.829	0.798	0.971	1.023	0.747	0.831	0.975	1.004	0.724	0.838	0.976
T + W + P + Rh + Sh	1.081	0.808	0.809	0.972	**1.009**	**0.728**	**0.836**	**0.976**	1.060	0.785	0.817	0.973

TABLE 8.48

Statistical Indices of SMO–SVR Models with Combinations of Input Parameters for Bangalore Station (Testing Data)

Input Combination	SMO–SVR Polynomial Kernel			SMO–SVR RBF Kernel			SMO–SVR PUK Kernel		
	RMSE (mm)	MAE (mm)	CC	NSE	RMSE (mm)	MAE (mm)	CC	NSE	RMSE (mm)	MAE (mm)	CC	NSE
T	1.519	1.059	0.571	0.943	1.485	0.997	0.600	0.946	1.481	1.007	0.600	0.946
T + W	1.543	1.080	0.563	0.941	1.545	1.049	0.569	0.941	1.595	1.116	0.547	0.937
T + W+ P	1.519	1.028	0.582	0.943	1.497	0.984	0.610	0.945	1.520	1.017	0.593	0.943
T + W + P + Rh	1.379	0.879	0.673	0.953	1.390	0.883	0.668	0.952	1.395	0.881	0.668	0.952
T + W + P + Rh + Sh	1.431	0.912	0.651	0.949	1.382	0.887	0.681	0.959	1.389	0.888	0.671	0.952

seasonal atmospheric variations. The sparseness in data recorded was found more with Bangalore data than Bajpe data. The individual SVR model has captured these seasonal variations to its limitations.

The computed results confirm the RBF's superiority in modeled estimations in both training and testing phases for the varied climatic stations. Among the other two kernel estimations PUK were close to RBF estimations. The polynomial showed the most contrast in results with RBF in both training and testing modeled estimations for Bangalore station.

Figures 8.51–8.53 represent SVR polynomial, RBF, and PUK estimations of pan evaporation over observed pan evaporation for Bajpe and Bangalore stations, respectively. Temporal variation of estimated pan evaporation and its comparison with observed pan evaporation for RBF kernel are in better agreement than the other two. RBF captured these variations better in the training and testing phases. At higher pan evaporation, value estimation almost matches the observed pan evaporation. Neglecting the seasonal irregularity, RBF estimations are closer to observed pan evaporation values.

Discussing these time series plots station-wise, polynomial kernel estimations are poor for the both the stations. It has failed to match the observed points as seen with the modeled results in both the phases as well as for both stations. As previously discussed, RBF kernel estimations for Bajpe station in testing is better than training. In plots, low and middle range pan evaporation values at testing phase are closer to observed values. But is exhibited in the results trend of the Bangalore station, as RBF estimated values are closer to observed values in the training phase than testing phase. PUK kernel

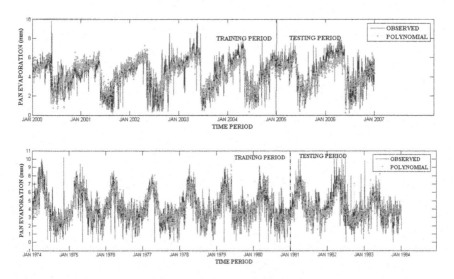

FIGURE 8.51
Pan evaporation estimation using polynomial kernel for Bajpe (top) and Bangalore station (bottom).

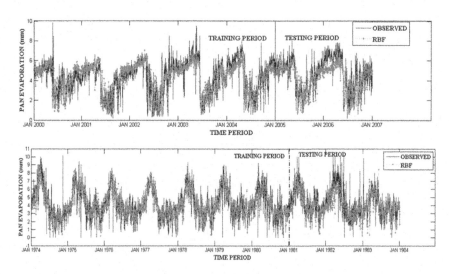

FIGURE 8.52
Pan evaporation estimation using RBF kernel for Bajpe (top) and Bangalore station (bottom).

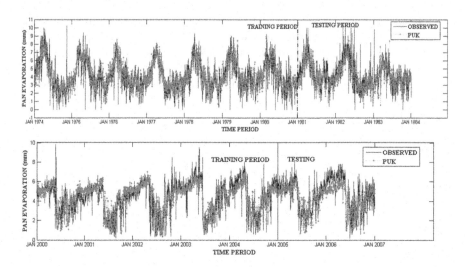

FIGURE 8.53
Pan evaporation estimation using PUK kernel for Bajpe (top) and Bangalore station (bottom).

estimation pattern was similar to RBF for the estimated results of both the stations, except during few periods where there are noisy data.

In the summer season (April–June), the modeled pan evaporation date are found underestimated compared to observed values during the training and testing periods. During the rainy season (July–September), modeled pan evaporation is in close agreement or slightly overestimated than observed values. In general, this trend is found to be uniform for all the kernel-based functions, such as RBF, PUK, and polynomial under SVR. However, an edge of superiority for RBF over the other two kernel functions of PUK and polynomial in modeling pan evaporation for Bajpe station was observed.

Similar patterns of performance were observed in Bangalore station for the training period, but kernel estimations were found to have some gaps between modeled and observed values.

Figure 8.54 represents the time series hydrograph plots of estimated pan evaporation using SVR-polynomial kernel along with the observed pan evaporation for the testing periods of 2005–06 and 2006–07, respectively, at station Bajpe. As seen in these plots, polynomial estimations exhibit certain difference with the observed pan evaporation values for almost entire testing periods. Polynomial kernel estimated values are below par for the period 2006–07 particularly during the crucial periods, such as January–May, when evaporation is greater. Overall, the estimates computed by polynomial kernels were underestimated compared with the observed values for the majority of the testing periods mentioned.

Figure 8.55 shows SVR-RBF kernel estimated pan evaporation against the observed pan evaporation time series hydrograph plots for the testing periods of 2005–06 and 2006–07, respectively, at station Bajpe. For the seasonal periods where the observed pan evaporation is greater, RBF provided fairly accurate estimations particularly for the period 2005–06. Overall, the estimated values nearly trace the observed values. But for the time series plot of 2006–07, initial pan evaporation values ranging between 3 and 7, RBF kernel

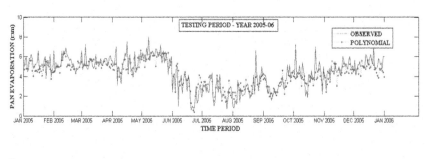

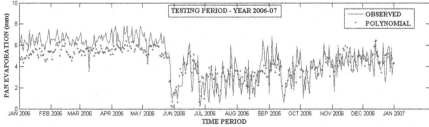

FIGURE 8.54
Daily pan evaporation estimation using SVR–Polynomial kernel for testing period 2005–06 and 2006–07 of Bajpe station.

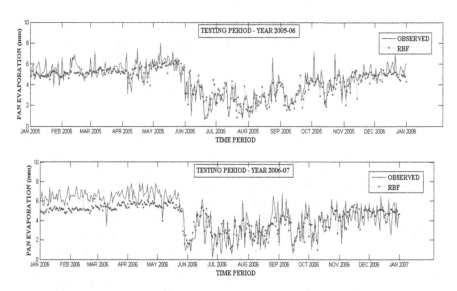

FIGURE 8.55
Daily pan evaporation estimation using SVR–RBF kernel for testing period 2005–06 and 2006–07 of Bajpe station.

underestimates the estimated values over observed pan evaporation values during the period January 2006–June 2006. However, there is improved accuracy in the estimations for the other ranges of pan evaporation values.

Figure 8.56 shows the SVR-PUK kernel estimated pan evaporation along with the observed pan evaporation time series plots for the testing periods of 2005–06 and 2006–07, respectively, at station Bajpe. The PUK estimation was closer to the one provided by RBF. As seen previously in RBF kernel plots, PUK also failed to accurately estimate the observed pan evaporation values during the crucial period of January 2006–June 2006. Furthermore, the PUK estimation was better for low-range pan evaporation values than peak values of pan evaporation.

Figure 8.57 displays time series hydrograph plots of estimated pan evaporation using SVR-polynomial kernel along with the observed pan evaporation for the testing periods of 1982–83 and 1983–84, respectively, at station Bangalore. The testing phase estimations provided by all the three SVR kernel estimations for the Bangalore station are not as satisfactory as Bajpe. However, improved accuracy in estimations for the crucial periods such as January–May, which are very important and essential for various water resources field applications, has been observed in these plots. Taking these results in regard to testing periods into consideration, the performance trend is quite satisfactory for the period 1983–84 compared with 1982–83. This is because nonlinearity in the observed data was the minimum for the period 1983–84 compared with the period 1982–83.

Figure 8.58 shows SVR-RBF kernel estimated pan evaporation compared with the observed pan evaporation time series hydrograph plots for the

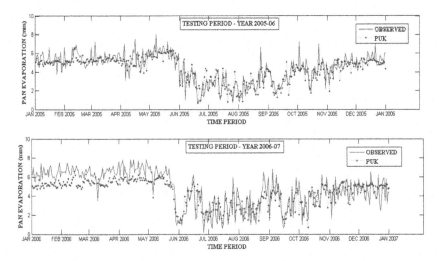

FIGURE 8.56
Daily pan evaporation estimation using SVR–PUK kernel for testing period 2005–06 and 2006–07 of Bajpe station.

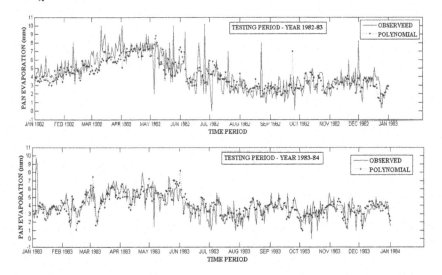

FIGURE 8.57
Daily pan evaporation estimation using SVR–Polynomial kernel for testing period 1982–83 and 1983–84 of Bangalore station.

testing periods of 1982–83 and 1983–84, respectively, at station Bangalore. RBF estimates were better in comparison to the remaining two kernel estimates for the testing periods. The modeled results are almost underestimated compared to observed values for the majority of these testing periods. The estimates match the observed pan evaporation for the period 1983–84. The peak value in the periods of 1982–83 was not reliably captured by RBF estimations.

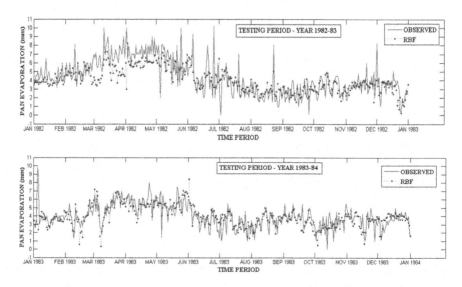

FIGURE 8.58
Daily pan evaporation estimation using SVR–RBF kernel for testing period 1982–83 and 1983–84 of Bangalore station.

Figure 8.59 shows the SVR-PUK kernel estimated pan evaporation along with the observed pan evaporation time series plots for the testing periods of 1982–83 and 1983–84, respectively, at station Bangalore. Performance trend is comparable to the RBF. The limitations of SVR kernels in capturing highly nonlinear pan evaporation trends is once again displayed in the PUK estimation plots. The PUK estimates exhibit small gaps in observed data for middle- and low-range pan evaporation values.

In the scatter plots shown in Figure 8.60, the pan evaporation estimates of RBF seem to be less scattered. The RBF model's estimates are closer to the ideal line than those of the other models, especially for the mean and peak pan evaporation values as observed in both the stations' estimates. Although RBF estimates are closer to observed pan values, polynomial kernel values with lower range are very scattered from the ideal line with Bangalore station estimates than Bajpe station estimates. The PUK estimates compete better with RBF and are closer to the ideal line. It is also seen in these plots that middle range pan evaporation estimates are much closer to observed values and are less scattered from the dividing line.

As far as kernel estimates are concerned, the Bajpe station estimates are almost overestimated compared to the observed values, whereas Bangalore estimations are underestimated compared to the observed values. This indicates that the variations observed in the statistical range of attributes of these

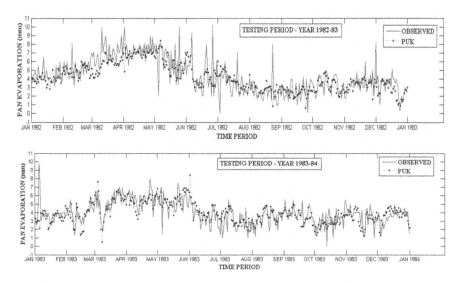

FIGURE 8.59
Daily pan evaporation estimation using SVR–PUK kernel for testing period 1982–83 and 1983–84 of Bangalore station.

stations have influenced the model estimations. The ranges in the attributes of the Bangalore station were especially varied.

Due to the larger data set containing daily pan evaporation data, these scatter plots seem to be too clumsy to rank superiority of any kernel-based function performance. Most of the model results for training of both the stations are found with similar performance levels, except polynomial kernel estimations (particularly for low and very high range values). Also, for both the stations, performance trends are similar as demonstrated in Figure 8.60.

It may be concluded that all the kernel estimations are closer to observed values in the training phase. The consistency of model performance in dealing with two contrasting statistical data sets of meteorological attributes was superior and flexible with RBF kernel-based estimations, which also demonstrated robustness capability.

Figures 8.61 and 8.62 represent scatter plots of testing data of Bajpe and Bangalore. Scatter plots of the training period, which were clumsy, could not establish the superiority rank of the kernel estimations. In testing period plots, the data points are limited. It becomes possible to distinguish the performance of kernel estimations with each other.

Regarding the Bajpe station scatter plot of period 2005–06 in Figure 8.61, it is observed that the scattering of estimated points from the ideal line is due to the polynomial kernel estimated values, most of which were found to be underestimated. RBF kernel estimates are close to the ideal line with better

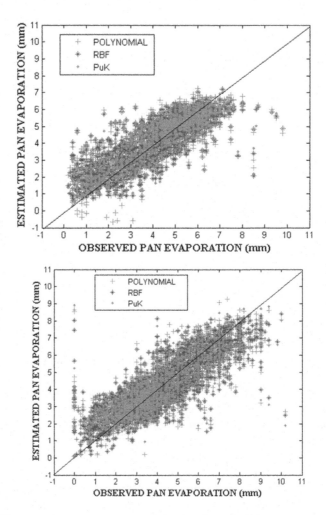

FIGURE 8.60
Scatter plot between SVR kernels estimated and observed daily pan evaporation training data for Bajpe (top) and Bangalore (bottom) stations.

accuracy than PUK kernel. PUK estimates are less scattered compared to polynomial kernel, but the observed values were underestimated. For middle range pan evaporation values, all three kernels estimated values are in better agreement with observed values on average. For the range six to nine, polynomial and PUK estimated values scatter widely from the ideal line.

The scatter plot for the testing period 2006–07 of Bajpe station shows the varied accuracy of these three kernel estimations. The polynomial and PUK estimated values deviated more from the observed points than RBF.

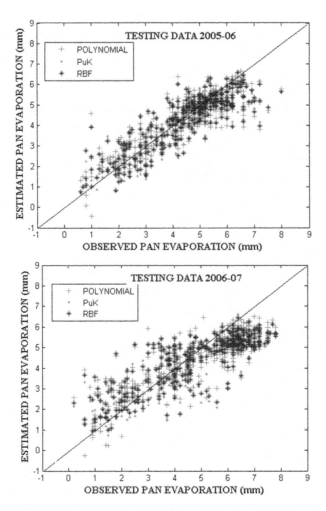

FIGURE 8.61
Scatter plot between estimated and observed daily pan evaporation testing data for Bajpe station.

In Figure 8.62, for the testing period of 1982–83, estimated values of polynomial kernel are more scattered for most of the range of pan evaporation. RBF estimates are quite good for the range two to four. But as the range exceeds four, the accuracy decreases and estimated values show more deviation with observed data. For the testing period 1983–84, the estimated values are almost balanced and show equal spacing on either side with respect to the ideal line. This indicates that at some periods, the model has overestimated the observed values and vice versa.

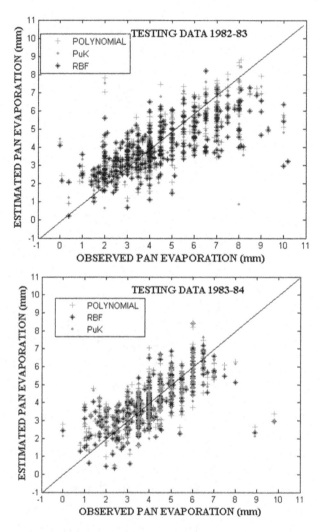

FIGURE 8.62
Scatter plot between estimated and observed daily pan evaporation testing data for Bangalore station.

8.9.3 Genetic Programming in Sea Wave Height Forecasting

The parameters for GP, such as population size, number of generations, mutation frequency and crossover frequencies are decided based upon previous studies.

 Population size – 500
 Number of generations 300

Mutation frequency – 90%

Crossover frequency – 50%

It is understood that these values have been referred from previous research, and the performance of the GP in the mentioned software (DISCIPULUS) or any other may or may not differ from the optimization values that are cited in this work. The performance evaluation is carried out using R^2, R, Bias, and Scatter Index evaluators. The formulas for the indices are already given. Although the basic structure of the error evaluation remains the same, there are certain probabilistic indices which give results in terms of overestimation and underestimation.

Genetic Programming with Varying Percentages of Training, Validation and Testing Data

40-20-40 ratio

For the given data, the data division was carried out in the following manner.

Training data – 40%
Validation data – 20%
Testing data – 40%

The following were the performance indices.

40-30-30 ratio

For the given data, the data division was carried out in the following manner.

Training data – 40%
Validation data – 30%
Testing data – 30%

The following were the performance indices.

50-25-25 ratio

TABLE 8.49

Performance Index for 40-20-40

% Data Distribution	R^2	BIAS	SI
40-20-40	0.463	1.10	0.281

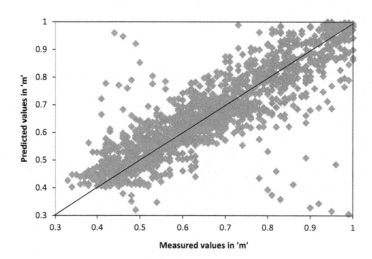

FIGURE 8.63
GP with 40-20-40 as training, validation and testing data sets respectively.

TABLE 8.50

Performance Index for 40-30-30

% Data Distribution	R²	BIAS	SI
40-30-30	0.800	0.98	0.201

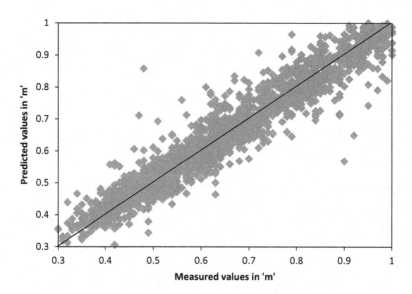

FIGURE 8.64
GP with 40-30-30 as training, validation and testing data sets respectively.

For the given data, the data division was carried out in the following manner.

Training data – 50%
Validation data – 25%
Testing data – 25%

The following were the performance indices.

60-20-20 ratio

For the given data, the data division was carried out in the following manner.

Training data – 60%
Validation data – 20%
Testing data – 20%

The following were the performance indices.

70-15-15 ratio

TABLE 8.51

Performance Index for 50-25-25

% Data Distribution	R²	BIAS	SI
50-25-25	0.561	0.998	0.143

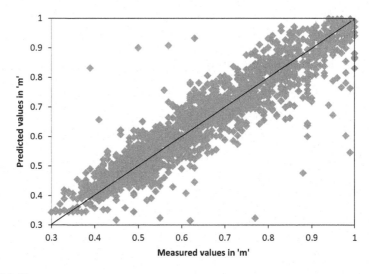

FIGURE 8.65
GP with 50-25-25 as training, validation and testing data sets respectively.

TABLE 8.52

Performance Index for 60-20-20

% Data Distribution	R^2	BIAS	SI
60-20-20	0.734	0.999	0.121

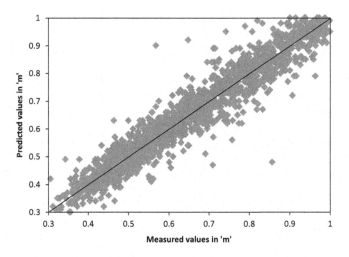

FIGURE 8.66
GP with 60-20-20 as training, validation and testing data sets respectively.

For the given data, the data division was carried out in the following manner.

Training data – 70%
Validation data – 15%
Testing data – 15%

The following were the performance indices.

80-10-10 ratio

For the given data, the data division was carried out in the following manner.

Training data – 80%
Validation data – 10%
Testing data – 10%

The following were the performance indices.

TABLE 8.53

Performance Index for 70-15-15

% Data Distribution	R^2	BIAS	SI
70-15-15	0.921	1.00005	0.038

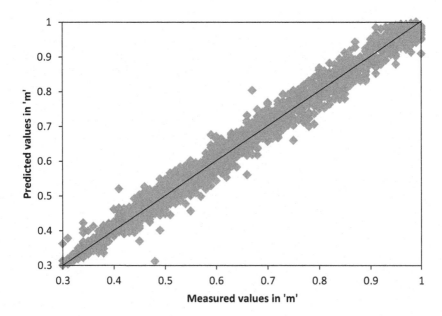

FIGURE 8.67
GP with 70-15-15 as training, validation and testing data sets respectively.

TABLE 8.54

Performance Index for 80-10-10

% Data Distribution	R²	BIAS	SI
80-10-10	0.921	1.00007	0.041

TABLE 8.55

GP with Varying Data Distribution Percentages

% Data Distribution	R²	BIAS	SI
40-20-40	0.463	1.10	0.281
40-30-30	0.800	0.98	0.201
50-25-25	0.749	0.998	0.143
60-20-20	0.857	0.999	0.121
70-15-15	**0.96**	**1.00005**	**0.038**
80-10-10	0.96	1.00007	0.041

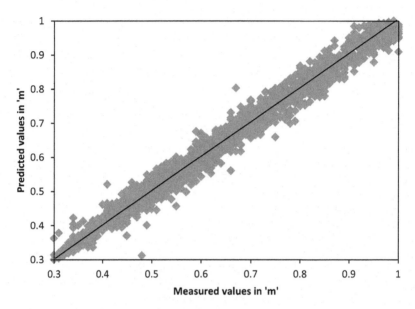

FIGURE 8.68
GP with 80-10-10 as training, validation, and testing data sets respectively.

Bibliography

Achim, D., Ghotb, F., Mcmanus, K.J. 2007. Prediction of water pipe asset life using neural networks, *Journal of Infrastructure Systems*, 13(1), 26–30.

Affandi, A.K. and Watanabe, K. 2007. Daily groundwater level fluctuation forecasting using soft computing technique, *Journal of Nature and Science*, 5(2), 1–10.

Affandi, A.K., Watanabe, K., and Tirtomihardjo, H. 2007. Application of an artificial neural network to estimate groundwater level fluctuation, *Journal of Spatial Hydrology*, 7(2), 17–32.

Agyare, W.A., Park, S.J., and Vlek, P.L. 2007. Artificial neural network estimation of saturated hydraulic conductivity, *Vadose Zone Journal*, 6(2), 423–431.

Aijun, A.N., Shan, N., Chan, C., Cercone, N., Ziarko, W. 1996. Discovering rules for water demand prediction: AN enhanced Rough-set approach, *Engineering Applications of Artificial Intelligence*, 9(6), 645–653.

Akkurt, S., Tayfur, G., Can, S. 2004. Fuzzy logic model for the prediction of cement compressive strength, *Cement and Concrete Research*, 34, 1429–1433.

Altunkaynak, A., Ozger, M., Cakmakci, M. 2005. Water consumption prediction of Istanbul city by using Fuzzy logic approach, *Water Resources Management* 19, 641–654.

Amirmojahedi, M., Mohammadi, K., Shamshirband, S., Seyed Danesh, A., Mostafaeipour, A., and Kamsin, A. 2016. A hybrid computational intelligence method for predicting dew point temperature. *Environmental Earth Sciences*, 75(5), 415. doi:10.1007/s12665-015-5135-7.

Arshad, R., Sayyad, G., Mosaddeghi, M., and Gharabaghi, B. 2013. Predicting saturated hydraulic conductivity by artificial intelligence and regression models, *ISRN Soil Science*, Hindawi Publishing Corporation, 2013, 1–8. doi: org/10.1155/2013/308159.

ASCE Task Committee. 2000a. Artificial neural networks in hydrology-I: preliminary concepts, *Journal of Hydrologic Engineering, ASCE*, 5(2), 115–123.

ASCE Task Committee. 2000b. Artificial neural networks in hydrology-II: hydrologic applications, *Journal of Hydrologic Engineering, ASCE*, 5(2), 124–137.

Aytek, A. and Kisi, O. 2008. A genetic programming approach to suspended sediment modeling. *Journal of Hydrology* 351, 288–298.

Babel, M. and Shinde, R. 2011. Identifying prominent explanatory variables for water demand prediction using artificial neural network: a case of Bangkok, *Water Resource Management* 25, 1653–1676.

Banzhaf, W., Nordin, P., Keller, R. E., and Francone, F.D. 1998. *Genetic Programming: An Introduction: On the Automatic Evolution of Computer Programs and Its Applications*, Morgan Kaufmann.

Besaw, L.E., Rizzo, D.M., Bierman, P.R., and Hackett, W.R. 2010. Advances in ungauged stream flow prediction using artificial neural networks, *Journal of Hydrology*, 386(2010), 27–37.

Bilgili, M. and Sahin, B. 2009. Prediction of long-term monthly temperature and rainfall in Turkey. *Energy Sources, Part A: Recovery, Utilization, and Environmental Effects*, 32(1), 60–71.

Chen, L., and Chen, C., Pan, Y. 2010. Groundwater level prediction using SOM-RBFN multisite model. *Journal of Hydrologic Engineering, ASCE*, 15(8), 624–631.

Cigizoğlu, H.K. 2005. Application of generalized regression neural networks to intermittent flow forecasting and estimation. *Journal of Hydrologic Engineering, ASCE*, 10(4), 336–341.

Cigizoğlu, H.K. and Kisi, O. 2005. Flow prediction by three back propagation techniques using k-fold partitioning of neural network training data, *Nordic Hydrology*, 36(1), 49–64.

Cladera, A. and Mari, A.R. 2004. Shear design procedure for reinforced normal and high-strength concrete beams using artificial neural networks Part I: beams without stirrups, *Engineering Structures*, 26, 917–926.

Cladera, A. and Mari, A.R. 2004. Shear design procedure for reinforced normal and high-strength concrete beams using artificial neural networks Part II: beams with stirrups, *Engineering Structures*, 26, 927–936.

Coppola, E., Poulton, M., Charles, E., Dustman, J. and Szidarovszky, F. 2003. Application of artificial neural network s to complex groundwater management problems, *Journal of Natural Resources Research*, 12(4), 303–320.

Coppola, E., Rana, A., Poulton, M., Szidarovszky, F., Uhl, V. 2005. A neural network model for predicting aquifer water level elevations. *Journal of Ground Water*, 43(2), 231–241.

Coulibaly, P., Anctil, F., Aravena, R., Bobee, B. 2001. Artificial neural network modeling of water table depth fluctuations. *Water Resources Research*, 37(4), 885–896.

Coulibalya, P. and Baldwin, C.K. 2005. Nonstationary hydrological time series forecasting using nonlinear dynamic methods, *Journal of Hydrology* 307(2005), 164–174.

Coulibalya, P., Anctil, F., Bobée, B. 2000. Daily reservoir inflow forecasting using artificial neural networks with stopped training approach, *Journal of Hydrology*, 230(2000), 244–257.

Cox, D.T., Tissot, P., and Michaud, P. 2002. Water level observations and short-term predictions including meteorological events for entrance of Galveston Bay, Texas. Journal of Waterway, *Coastal and Ocean Engineering*, 128(1), 21–29.

Cui, L. and Sheng, D. 2005. Genetic algorithms in probabilistic finite element analysis of geotechnical problems, *Computers and Geotechnics*, 32, 555–563.

Daneshmand, H., Tavousi, T., Khosravi, M., and Tavakoli, S. 2012. Modeling minimum temperature using adaptive neuro-fuzzy inference system based on spectral analysis of climate indices: a case study in Iran, *Journal of the Saudi Society of Agricultural Sciences*, 14(1), 33–40.

Das, S., Manna, B. and Baidya, D.K. 2011. Prediction of dynamic Soil-Pile Interaction under coupled vibration using artificial neural network approach, *Geo Frontiers*, pp. 1–10.

Dawson, C.W. and Wilby, R. 1998. An artificial neural network approach to rainfall runoff modeling, *Hydrological Sciences Journal*, 43(1), 47–66.

Deka, P.C. and Prahlada, R. 2012. Discrete wavelet neural network approach in significant wave height forecasting for multistep lead time, *Ocean Engineering*, 43, 32–42.

Deo, M.C. and Chaudhari, G. 1998. Tide prediction using neural networks, *Computer-Aided Civil and Infrastructure Engineering*, 13(2), 113–120.

Deo, M.C. and Naidu, C.S. 1999. Real time wave forecasting using neural network. *Ocean Engineering*, 35, 191–203.

Deswal, S. and Pal, M. 2008. Artificial neural network based modeling of evaporation losses in reservoirs, *International Journal of Mathematical, Physical and Engineering Sciences*, 39(3), 177–181.

Dias, W.P.S. and Pooliyadda, S.P. 2001. Neural networks for predicting properties of concretes with admixtures, *Construction and Building Materials*, 15, 371–379.

Dombayc, Ö.A. and Gölcü, M. 2009. Daily means ambient temperature prediction using artificial neural network method: a case study of Turkey, *Renewable Energy*, 34(4), 1158–1161.

Drucker, H., Wu, D., and Vapnik, V.N. 1999. Support vector machines for spam categorization, *IEEE Transactions on Neural Networks*, 10, 1048–1054.

Durga Rao, K.H.V. 2005. Multicriteria spatial decision analysis for forecasting urban water requirement: a case study of Dehradun city, India. *Landscape and Urban Planning*, 71, 163–174.

Erfani, M. and Farsangi, E.N. 2010. Fuzzy neural network utilization in prediction of compressive strength of slag-cement based mortars, *Australian Journal of Basic and Applied Sciences*, 4(10), 4962–4970.

Falco, D., Cioppa, A.D., and Tarantino, E. 2005. A genetic programming system for time series prediction and its application to El Nino forecast, *Advances in Soft Computing*, 32(2005), 151–162.

Feng, Y., Cui, N., Zhao, L., Hu, X., and Gong, D. 2016. Comparison of ELM, GANN, WNN and empirical models for estimating reference evapotranspiration in humid region of Southwest China, *Journal of Hydrology*, 536, 376–383. doi:10.1016/j.jhydrol.2016.02.053.

Firat, M., Tarun, M., Yurdusev, M. 2010. Comparative analysis of neural network technique for predicting water consumption time period, *Journal of Hydrology*, 384, 46–51.

Firat, M., Yurdusev, M., Turan, M. 2009. Evaluation of artificial neural network techniques for municipal water consumption modeling, *Water Resource Management*, 23, 617–632.

Gat, Y.L. and Eisenbeis P. 2000. Using maintenance records to forecast failures in water networks, *Journal of Urban Water*, 2, 173–181.

Gaur, S. and Deo, M.C. 2008. Real time wave forecasting using genetic programming. *Ocean Engineering* 35, 1166–1172.

Ghannadpour, S.F., Noori, S., and Tavakkoli-Moghaddam, R. 2013. Multiobjective dynamic vehicle routing problem with fuzzy travel times and customers satisfaction in supply chain management, *IEEE Transactions on Engineering Management*, 60(4), 777–790.

Ghannadpour, S.F., Noori, S., Tavakkoli-Moghaddam, R., Ghoseiri, K. 2014. A multiobjective dynamic vehicle routing problem with fuzzy time windows: model, solution and application, *Applied Soft Computing*, 14, 504–527.

Graham, L.D., Forbes, D.R., and Smith, S.D. 2006. Modeling the ready mixed concrete delivery system with neural networks, *Automation in Construction*, 15, 656–663.

Graham, L.D., Forbes, D.R., and Smith, S.D. 2006. Modeling the ready mixed concrete delivery system with neural networks, *Journal of Automation in Construction*, 15, 656–663.

Guang, N.H. and Zong, W.J. 2000. Prediction of compressive strength of concrete by neural networks, *Journal of Cement and Concrete Research*, 30, 1245–1250.

Hadi, M.N.S. 2003. Neural network applications in concrete structures, *Computers and Structures*, 81(6), 373–381.

Herrera, M., Torgo, L., Izquierdo, J., Garcia, R. 2010. Predictive models for forecasting hourly urban water demand, *Journal of Hydrology* 387, 141–150.

Hildebrandt, G., Bazartseren B., and Holz, K.-P. 2003. Short-term water level prediction using neural networks and neuro-fuzzy approach, *Neurocomputing*, 55, 439–450.

Ho, C.I., Lin, M.D., Lo, S.L. 2009. Use of GIS- based hybrid artificial neural network to prioritize the order of pipe replacement in a water distribution network. *Environmental Monitoring and Assessment*, 166, 177–189.

Hongwei, Z., Xuehua, Z., Bao, Z. 2009. System dynamic approach to urban water demand forecasting: a case study of Tianjin. Tianjin University and Springer-Verlag 15, 70–74.

hua Tian, C., Xiao, J., Huang, J., Albertao, F. 2011. Pipe failure prediction, IBM Research –*China Beijing*, 100193, 121–125.

Huang, G.-B., Zhu, Q.-Y., and Siew, C.-K. 2006. Extreme learning machine: theory and applications. *Neurocomputing*, 70(1–3), 489–501. doi:10.1016/j.neucom.2005.12.126.

Jaber, A. and Fawzia, A.-R. 2006. Periodic behavior of groundwater level fluctuations in residential areas, *Journal of Hydrology*, 328, 677–684.

Jafar, R., Shahrour, I., and Juran., I. 2010. Application of artificial neural networks to model the failure of urban water mains, *Mathematical and Computer Modeling*, 51, 1170–1180.

Jain, A., Varshney, K., Joshi, U. 2001. Short term water demand forecasting modeling at IIT Kanpur using artificial neural networks, *Water Resources Management*, 15, 299–321.

Jain, P. and Deo, M.C. 2006. Neural networks in ocean engineering, *International Journal of Ships and Offshore Structures*, Cambridge, UK, 1(1), 25–35.

Jain, P., and Deo, M.C. 2008. Artificial intelligence tools to forecast ocean waves in real time. *The Open Ocean Engineering Journal*, 1, 13–20.

Jain, S.K., Nayak, P.C., and Sudheer, K.P. 2008. Models for estimating evapotranspiration using artificial neural networks and their physical interpretation, *Hydrological Processes*, 22, 2225–2234. doi: 10.1002/hyp.

Jia, B., Feng, W., and Zhu, M. 2016. Obstacle detection in single images with deep neural networks, *Signal, Image and Video Processing*, 10(6), 1033–1040. doi: 10.1007/s11760-015-0855-4.

Jiang, Y. and Shao, M. 2014. Effects of soil structural properties on saturated hydraulic conductivity under different land-use types, *Soil Research*, 52(4), 340–348.

Jirků, V., Kodešová, R., Nikodem, A., Mühlhanselová, M., and Žigová, A. 2013. Temporal variability of structure and hydraulic properties of topsoil of three soil types, *Geoderma*, 204–205, 43–58.

Joorabchi, A., Zhang, H., and Bluemenstein, M. 2009 Application of artificial neural networks to groundwater dynamics in coastal aquifers, Journal of Coastal Research, Special Issue 50, 966–970.

Kalteh, M., et al. 2008. Rainfall-runoff modeling using artificial neural networks (ANNs): modeling and understanding, *Caspian Journal of Environmental Sciences*, 6(1), 53–58.

Kazeminezhad, M.H., Etemad-Shahidi, A., and Mousavi, S.J. 2005. Application of fuzzy inference system in the prediction of wave parameters, *Ocean Engineering*, 32, 1709–1725.

Kermani, Z. and Teshnehlab, M. 2008. Using adaptive neuro fuzzy inference system for hydrological time series prediction, *Applied Soft Computing*, 8, 928–936.

Kim, H., Hwang, S., Shin, H. 2001. A neuro-genetic approach for daily water demand forecasting, *KSCE Journal of Civil Engineering*, 5(3), 281–288.

Kim, S., Shiri, J., and Kisi, O. 2012. Pan evaporation modeling using neural computing approach for different climatic zones, *Water Resource Management*, 26(11), 3231–3249. doi: 10.1007/s11269-012-0069-2.

Kim, S., Singh, V.P., Lee, C.-J., and Seo, Y. 2015. Modeling the physical dynamics of daily dew point temperature using soft computing techniques. *The KSCE Journal of Civil Engineering*, 19(6), 1930–1940. doi: 10.1007/s12205-014-1197-4.

Kişi, O. 2006b. Generalized regression neural networks for evapotranspiration modeling, *Journal of Hydrological Science* 51(6), 1092–1104.

Kisi, O. and Shiri, J. 2014. Prediction of long-term monthly air temperature using geographical inputs, *International Journal of Climatology*, 34(1), 179–186.

Kisi, O., Kim, S., and Shiri, J. 2013. Estimation of dew point temperature using neuro-fuzzy and neural network techniques, *Theoretical and Applied Climatology*, 114(3–4), 365–373. doi:10.1007/s00704-013-0845-9.

Kleiner, Y. and Rajani, B. 2002. Forecasting variations and trends in water main breaks. *Journal of Infrastructure Systems*, 8, 122–131.

Koesdwiady, A., Soua, R., and Karray, F. 2016. Improving traffic flow prediction with weather information in connected cars: a deep learning approach, *IEEE Transactions on Vehicular Technology*, 65(12), 9508–9517.

Krishna, B., Satyaji Rao, Y.R., and Vijaya, T. 2008. Modeling groundwater levels in an urban coastal aquifer using artificial neural networks, *Hydrological Processes*, 228, 1180–1188.

Kumar, A.M., Jain, A. 2007. Hybrid neural network models for hydrologic time series forecasting, Elsevier, *Applied Soft Computing*, 7, 585–592.

Kumar, K., Parida, M., and Katiyar, V. 2015. Short term traffic flow prediction in heterogeneous condition using artificial neural network, *Transport*, 30(4), 397–405.

Kumar, P. 2012. Minimum weekly temperature forecasting using ANFIS, *Computer Engineering and Intelligent Systems*, 3(5), 1–5.

Lee, T.-L. and Jeng, D.S. 2002. Application of artificial neural networks in tide forecasting, *Ocean Engineering*, 29, 1003–1022.

Lee, T.-L., Tsai, C.-P., Jeng, D.S., and Shieh, R.J. 2002. Neural network for the prediction and supplement of tidal record in Taichung, Taiwan, *Advances in Engineering Software*, 33, 329–338.

Liang, S.X., Li, M.C., and Sun, Z.C. 2007. Prediction models for tidal level including strong meteorologic effects using a neural network, *Ocean Engineering*, 35, 666–675.

Liu, M., Qie, Z., Wu, X., Dong, W. and Zheng, H. 2008. Model optimization of load-bearing capacity of macadam pile composite foundation based on genetic algorithms, IEEE Control and Decision Conference, Yantai, pp. 3903–3907.

Londhe, S. and Panchang, V. 2006. One-day Wave forecast based on artificial neural network, *Journal of Atmospheric and Oceanic Technology*, 23(11), 593–603.

Lopez-Garcia, P., Onieva, E., Osaba, E., Masegosa, A., and Perallos, A. 2016. A hybrid method for short-term traffic congestion forecasting using genetic algorithms and cross entropy, *IEEE Transactions on Intelligent Transportation Systems*, 17(2), 557–569.

Lu, W.Z. and Wang, W.J. 2005. Potential assessment of the "support vector machine" method in forecasting ambient air pollutant trends, *Chemosphere*, 59(5), 693–701.

Makarynskaa, D. and Makarynskyy, O. 2008. Predicting sea-level variations at the Cocos (Keeling) Islands with artificial neural networks, *Computers & Geosciences* 34, 1910–1917.

Merdun, H., Çınar, Ö., Meral, R., and Apan, M. 2006. Comparison of artificial neural network and regression pedotransfer functions for prediction of soil water retention and saturated hydraulic conductivity, *Soil and Tillage Research*, 90, 108–116. doi: 10.1016/j.still.2005.08.011.

Moghaddamnia, A., Ghafari, M., Piri, J., Han, D. 2008. Evaporation estimation using support vector machines techniques, *World Academy of Science, Engineering and Technology*, 19(1), 14–22.

Mohammadi, K., Shamshirband, S., Motamedi, S., Petković, D., Hashim, R., and Gocic, M. 2015. Extreme learning machine based prediction of daily dew point temperature, *Computers and Electronics in Agriculture*, 117, 214–225. doi:10.1016/j.compag.2015.08.008.

Moosavi, V., Vafakhah, M., Shirmohammadi, B., Behnia, N. 2013. A wavelet-ANFIS hybrid model for groundwater level forecasting for different prediction periods, *Water Resources Management*, 27(5), 1301–1321.

More, S.B. and Deka, P.C. 2018. Estimation of saturated hydraulic conductivity using fuzzy neural network in a semi-arid basin scale for murum soils of India, *ISH Journal of Hydraulic Engineering*, 24, 140–146. doi: 10.1080/09715 010.2017.14004 08.

Motaghian, H.R. and Mohammadi, J. 2011. Spatial estimation of saturated hydraulic conductivity from terrain attributes using regression, kriging, and artificial neural networks, *Pedosphere* 21, 170–177. doi: 10.1016/S1002-0160(11) 60115-X.

Naganna, S.R. and Deka, P.C. 2018. Variability of streambed hydraulic conductivity in an intermittent stream reach regulated by Vented Dams: a case study, *Journal of Hydrology*, 562, 477–491. doi: 10.1016/j.jhydr ol.2018.05.006.

Naganna, S.R., Deka, P.C., Ch, S., and Hansen, W.F. 2017. Factors influencing stream-bed hydraulic conductivity and their implications on stream–aquifer inter-action: a conceptual review, *Environmental Science and Pollution Research*, 24, 24765–24789. doi: 10.1007/s11356-017-0393-4.

Nasseri, M., Moeini, A., Tabesh, M. 2011. Forecasting monthly urban water demand using extended Kalman filter and Genetic programming, *Expert System with Applications*, 38, 7387–7395.

Nath, U.K. and Hazarika, P.J. 2011. Study of Pile cap lateral resistance using artificial neural networks, *International Journal of Computer Applications*, 21(1), 20–25.

Nayak, P.C., Satyaji, Y.R., and Sudheer, K.P. 2006. Groundwater level forecasting in a shallow aquifer using artificial neural network, *Journal of Water Resource Management* 20, 77–90.

Nourani, V. and Sayyah, F.M. 2012. Sensitivity analysis of the artificial neural net-work outputs in simulation of the evaporation process at different climato-logic regimes, *Advances in Engineering Software*, 47(1), 127–146. doi: 10.1016/j.advengsoft.2011.12.014.

Onieva, E., Godoy, J., Villagr, J., Milans, V., and Prez, J. 2013. On-line learning of a fuzzy controller fora precise vehicle cruise control system, *Expert Systems with Applications*, 40(4), 1046–1053. Doi: 10.1016/j.eswa.2012.08.036.

Oreta, A.W.C. and Kawashima, K. 2003. Neural network modeling of confined com-pressive strength and strain of circular concrete columns, *Journal of Structural Engineering*, 129(4), 554–561.

Ozcan, F., Atis, C.D., Karahan, O., Uncuog˘lu, E., and Tanyildizi, H. 2009. Comparison of artificial neural network and fuzzy logic models for prediction of long-term compressive strength of silica fume concrete, *Advances in Engineering Software*, 40, 856–863.

Ozcan, F., Atis, C.D., Karahan, O., Uncuoglu, E., Tanyildiz, H. 2009. Comparison of artificial neural network and fuzzy logic models for prediction of long-term compressive strength of silica fume concrete, *Advances in Engineering Software*, 40(9), 856–863.

Ozcan, F., Atis, C.D., Karahan, O., Uncuoglu, E., Tanyildizi, H. 2009. Comparison of artificial neural network and fuzzy logic models for prediction of long-term compressive strength of silica fume concrete, *Advances in Engineering Software*, 40, 856–863.

Ozger, M. 2010. Significant wave height forecasting using wavelet fuzzy logic approach. *Ocean Engineering*, 37, 1443–1451.

Pallavi, B., Prasad, R.K. and Singh, V.P. 2009. Forecasting of groundwater level in hard rock region using artificial neural network, *Journal of Environmental Geology*, 58, 1239–1246.

Parasuraman, K., Elshorbagy, A., and Si, B.C. 2006. Estimating saturated hydraulic conductivity in spatially variable fields using neural network ensembles, *Soil Science Society of America Journal*, 70(6), 1851.

Parasuraman, K., Elshorbagy, A., and Si, B.C. 2007. Estimating saturated hydraulic conductivity using genetic programming, *Soil Science Society of America Journal*, 71(6), 1676.

Patil, A.P., and Deka, P.C. 2016. An extreme learning machine approach for mod-eling evapotranspiration using extrinsic inputs, *Computers and Electronics in Agriculture*, 121, 385–392. doi:10.1016/j.compag.2016.01.016

Pelletier, G., Mailhot, A. and Villeneuve J.-P. 2003. Modeling water pipe breaks—three case studies, *Journal of Water Resources Planning and Management*, 129(2), 115–123.

Radhika, Y. and Shashi, M. 2009. Atmospheric temperature prediction using support vector machines, *International Journal of Computer Theory and Engineering*, 1(1), 1793–8201.

Raghavendra, N.S. and Deka, P.C. 2014. Support vector machine applications in the field of hydrology: a review, *Applied Soft Computing*, 19, 372–386. doi: 10.1016/j.asoc.2014.02.002.

Raghavendra, N.S. and Deka, P.C. 2014. Support vector machine applications in the field of hydrology: a review, *Applied Soft Computing*, 19, 372–386. doi: 10.1016/j.asoc.2014.02.002.

Raghuwanshi, N.S., Singh, R., and Reddy, L.S. 2006. Runoff and sediment yield modeling using artificial neural, *Journal of Hydrologic Engineering*, 11(1), 71–79.

Rogers, P.D. and Grigg, N.S. 2009. Failure assessment modeling to prioritize water pipe renewal: two case studies, *Journal of Infrastructure Systems*, 15, 162–171.

Sanad, A. and Saka, M. 2001. Prediction of ultimate shear strength of reinforced-concrete deep beams using neural networks, *Journal of Structural Engineering*, 127(7), 818–828.

Saridemir, M. 2009. Predicting the compressive strength of mortars containing metakaolin by artificial neural networks and fuzzy logic, *Advances in Engineering Software*, 40(9), 920–927.

Savic, A.D., Walters, A.G., Davidson, J.W. 1999. A genetic programming approach to rainfall-runoff modeling, *Water Resource Management*, 13, 219–231.

Sen, Z. and Altunkaynak, A. 2009. Fuzzy system modeling of drinking water consumption prediction, *Expert Systems with Applications*, 36, 11745–11752.

Shahin, M.A., Maier, H.R. and Jaksa, M.B. 2002. Predicting settlement of shallow foundations using neural networks, *Journal of Geotechnical and Geoenvironmental Engineering*, 128(9), 785–793.

Shank, D.B., McClendon, R.W., Paz, J., and Hoogenboom, G. 2008. Ensemble artificial neural networks for prediction of dew point temperature, *Applied Artificial Intelligence*, 22(6), 523–542. doi:10.1080/08839510802226785.

Shiri, J., Kim, S., and Kisi, O. 2014. Estimation of daily dew point temperature using genetic programming and neural networks approaches, *Hydrology Research*, 45(2), 165. doi:10.2166/nh.2013.229.

Smith, B.A., McClendon, R.W., and Hoogenboom, G. 2006. Improving air temperature prediction with artificial neural networks, *International Journal of Computational Intelligence*, 3, 179–186.

Smith, B.A., McClendon, R.W., Hoogenboom, G. 2009. Artificial neural networks for automated year-round temperature prediction, *Computers and Electronics in Agriculture*, 68, 52–61.

Sobhani, J. 2010. Prediction of the compressive strength of no-slump concrete: a comparative study of regression, neural network and ANFIS models, *Construction and Building Materials*, 24, 709–718.

Suttasupa, Y., Rungraungsilp, S., Pinyopan, S., Wungchusunti, P., and Chongstitvatana, P. 2011. A comparative study of linear encoding in genetic programming, Ninth International Conference on ICT and Knowledge, ISBN 978-1-4577-2162-5/11.

Tabari, H., Kisi, O., Ezani, A., Hosseinzadeh, T.P. 2012 SVM, ANFIS, regression and climate based models for reference evapotranspiration modeling using limited climatic data in a semi-arid highland environment, *Journal of Hydrology*, 444–445, 78–89. doi: 10.1016/j.jhydrol.2012.04.007.

Tabesh, M., Soltani, J., Farmani, R., Savic, D.A. 2009. Assessing pipe failure rate and mechanical reliability of water distribution networks using data driven modeling, *Journal of Hydroinformatics*, 11(1), 1–17.

Tang, J., Pan, Z., Fung, R.Y., and Lau, H. 2009. Vehicle routing problem with fuzzy time windows. *Fuzzy Sets and Systems*, 160(5), 683–695.

Topcu, I.B. and Saridemi, M. 2008. Prediction of compressive strength of concrete containing fly ash using artificial neural networks and fuzzy logic, *Computational Materials Science*, 41(3), 305–311.

Twarakavi, N.K.C., Šimůnek, J., and Schaap, M.G. 2009. Development of pedotransfer functions for estimation of soil hydraulic parameters using support vector machines, *Soil Science Society of America Journal*, 73(5), 1443. doi: 10.2136/sssaj 2008.0021.

Twarakavi, N.K.C., Šimůnek, J., Schaap, M.G. 2009. Development of pedotransfer functions for estimation of soil hydraulic parameters using support vector machines, *Soil Science Society of America Journal*, 73, 1443.

Vapnik, V.N. 1995. *The Nature of Statistical Learning Theory*, Vol 8, Springer, Verlag, doi: 10.1109/TNN.1997.641482.

Wu, C.L. and Chau, K.W. 2010. Data-driven models for monthly stream flow time series prediction, Elsevier, *Engineering Applications of Artificial Intelligence*, 23(2010), 1350–1367.

Xu, C.Y. and Singh, V.P. 2001. Evaluation and generalization of temperature-based methods for calculating evaporation, *Hydrological Processes*, 15, 305–319. doi: 10.1002/hyp.11.

Yeh, I.-C. 1998. Modeling of strength of high performance concrete using artificial neural networks, *Cement and Concrete Research*, 28(12), 1797–1808.

Young-Su, K. and Byung-Tak, K. 2006. Use of artificial neural networks in the prediction of liquefaction resistance of sands, *Journal of Geotechnical and Geoenvironmental Engineering*, 132(11), 1502–1504.

Yu, X., Xiong, S., He, Y., Wong, W., and Zhao, Y. 2016. Research on campus traffic congestion detection using BP neural network and Markov model, *Journal of Information Security and Applications*, 31, 54–60.

Yurdusev, A. and Firat, M. 2009. Adaptive neuro fuzzy inference system approach for municipal water consumption modeling, *Journal of Hydrology*, 265, 225–234.

Zarandi, M.F., Türksen, I.B., Sobhani, J., and Ramezanianpour, A.A. 2008. Fuzzy polynomial neural networks for approximation of the compressive strength of concrete, *Applied Soft Computing* 8, 488–498.

Zhang, X., Onieva, E., Perallos, A., Osaba, E., and Lee, V. 2014. Hierarchical fuzzy rule-based system optimized with genetic algorithms for short term traffic congestion prediction, *Transportation Research Part C: Emerging Technologies*, 43, 127–142.

Zhao, C., Shao, M., Jia, X., Nasir, M., and Zhang, C. 2016. Using pedotransfer functions to estimate soil hydraulic conductivity in the Loess Plateau of China. *Catena*, 143, 1–6. doi:10.1016/j.caten a.2016.03.037.

Zhou, S.L., McMahon, T.A., Walton, A. and Lewis, J. 2000. Forecasting daily urban water demand: a case study of Melbourne, *Journal of Hydrology*, 236, 153–164.

Zounemat-Kermani, M. 2012. Hourly predictive Levenberg–Marquardt ANN and multi linear regression models for predicting of dew point temperature, *Meteorology and Atmospheric Physics*, 117(3–4), 181–192. doi:10.1007/s00703-012-0192-x.

9

Conclusion and Future Scope of Work

Conclusion

Problems encountered in the real world are sometimes difficult to be solved analytically due to the huge amount of resources and time required for computation. For these problems, computational tools inspired by nature sometimes work very efficiently and effectively. Although these tools do not provide exact answers to the problems, they are quite handy for problems governed by imprecise and vague data. These tools provide the user with sufficient decision-making capabilities when dealing with uncertainty and vagueness. Such tools are classified under the umbrella of soft computing with the aim of building intelligent and wiser machines.

Soft computing, through the fusion of methodologies, supplement the conventional computational techniques with the human-like capabilities of reasoning, intuition, consciousness, wisdom, and adaptability to changing environments. Two important soft computing techniques viz., Artificial Neural Networks and Genetic Algorithms have been described, and their applications in the diversified field of civil engineering have been reviewed. This work shows that ANNs, with the capability of harnessing the learning abilities resembling a human brain, can be used to model complex nonlinear functional relationships without having to assume any prior functional relationship.

The use of ANN in the field of civil engineering has made solving the complicated problems of modeling and prediction of behavior much easier. ANNs, learning from historical data and capturing the subtle functional relationships which surpassed most empirical and statistical methods already in practice in the field of engineering. The ANN addresses the problems whose solutions require prior knowledge which can be explicitly derived through historical data or experimental observations. Thus, ANN can be used as a decision support tool for engineering problems whose solutions do not rely on a definite sets of rules or algorithms.

In the field of civil engineering, Genetic Algorithm (GA) has been instrumental as an optimization tool dealing with discrete variables and the arbitrary nature of constraints and objectives. The stochastic search mechanism of GA, inspired by natural evolution, encourages an engineer to choose from a pool of multiple design alternatives. The operators of evolution viz., mutation, crossover, and selection, refines the search and helps GA to bring forth the best design alternative satisfying the design objective. The heuristic nature of GA provides improvement in the conventional design procedures aiding the development of economical and time-saving design tools. The works cited here have highlighted the use of GA as an optimization tool which explores and exploits the solution space and aims to find an optimal solution to a problem.

Hybridization of soft computing techniques has given a promising future to the development of next generation intelligent systems. The hybridization of GA and ANN, discussed in the earlier chapters, focuses on the complementing nature of soft computing tools which encourages the user to derive the best from these techniques. Hybridization enriches the original procedures, covers up the limitations of individual techniques, and helps the user resolve new problems. The conjunctive use of the techniques leads to accurate and robust solutions in comparison to those derived from an individual technique. Most of the applications of GA are centered on optimizing the neural network architecture and neural network weights during back-propagation learning. The result of ANN-GA hybridization is very promising, wherein the modeling capabilities of neural networks to derive complex functional relationships and the optimization capabilities of GAs are integrated to develop computational tools which can surely be a step beyond human cognitive skills.

Soft Computing (SC) is an emerging field that consists of complementary elements of fuzzy logic, neural computing, evolutionary computation, machine learning, and probabilistic reasoning. Due to their strong learning and cognitive ability and reliable tolerance of uncertainty and imprecision, soft computing techniques have found wide applications.

Soft computing is likely to play an especially important role in science and engineering, but eventually its influence may extend much further. Soft computing represents a significant paradigm shift in the aims of computing; a shift which reflects the fact that the human mind, unlike present day computers, possesses a remarkable ability to store and process information which is pervasively imprecise, uncertain, and lacking in categoricity.

Intelligent systems and, hence, soft computing techniques are becoming more important as the power of computer processing devices increases and the cost is reduced. Intelligent systems are required to make complex decisions and choose the best outcome from many possibilities, using complex algorithms. This requires fast processing power and large storage space which has recently become available in recent years to many research centers, universities, and technical colleges at a very low cost.

With the power and the recognition of the Internet of Things (IoT) concept, the need for using soft computing techniques and building intelligent systems has become more important than ever. Today, most soft computing applications can be handled efficiently by low cost, super-fast microcontrollers.

Already we see the use of fuzzy logic, artificial neural networks, and expert systems in many everyday domestic appliances, such as washing machines, cookers, and refrigerators. Many industrial and commercial applications of soft computing are also in everyday use, and this is expected to grow within the next decade. It is the author's opinion that soft computing theory and techniques, as well as its applications will grow rapidly together with the use of IoT devices in future domestic, industrial, and commercial markets.

Due to a lot of uncertain factors and complicated influence factors in civil engineering, each project has its individual character and generality; function of expert system in the special links and cases is a notable effect. Over the past 20 years in the civil engineering field, development and application of the expert system have made a lot of achievements, mainly in project evaluation, diagnosis, decision-making and prediction, building design and optimization, the project management construction technology, road and bridge health detection, and so forth.

This book summarizes and introduces the intelligent technologies in civil engineering with recent research results and applications presented. All aspects of applications of the artificial intelligence technology in civil engineering were analyzed. On the basis of said research results, prospects for artificial intelligence technology in the civil engineering field and development trends were represented. Artificial intelligence can help inexperienced users solve engineering problems, help experienced users to improve work efficiency, and help teams share the expertise of each member. Artificial intelligence technology will change with each passing day, as the computer is applied more and more popularly; and it has a broad prospect in the civil engineering field.

We have summarized the main bio-inspired methods for supply chain management (SCM) system design and optimization. It is appropriate to note that swarm-based methods and artificial immune systems are not yet mature and, thus, are expected to gain more research interest. In the civil engineering field, in the present situation, the research and development of artificial intelligence is only just starting—so far failing to play its proper role. The combination of artificial intelligence technology, and object-oriented programming, and the Internet is the general trend of development in artificial intelligence technology. Artificial intelligence is in development for civil engineering in the following aspects.

To deepen the understanding of the problems of uncertainty and to seek the appropriate reasoning mechanism is the primary task. To develop practical artificial intelligence technology, only to be developed in the field of artificial intelligence technology, and the knowledge to have a thorough grasp.

According to the application requirements of civil engineering, practical engineering, the research and development of artificial intelligence technology in the civil engineering field have been carried out continually. Many questions in this area need artificial intelligence technology. Due to the characteristics of the civil engineering field, artificial intelligence technology has been used in many areas, such as building engineering, bridge engineering, Mathematical Problems in Engineering geotechnical engineering, underground engineering, road engineering, geological exploration and structure of health detection, and so forth.

With the development of artificial intelligence technology, some early artificial intelligence technology needs enhancement and improvement in the areas of knowledge, reasoning mechanism, man–machine interface optimization, and so forth.

Many single functions of an artificial intelligent system can be integrated as a comprehensive system of artificial intelligence. Also, this system can be expanded to solve the question ability related to different problem types.

Artificial intelligence technology was used to test the reliability and give full play to the role that was assigned to it. In the commercialization of artificial intelligence technology, there are many successful examples that have brought considerable benefits to both private and social enterprises.

Script Files

- **Support Vector Machine (SVM)**

```
%% clear workspace and command window
clear;
clc;
close all;

%% prepare dataset
Input_Data = xlsread(filename);
Output_Data = xlsread(filename);

%% Normalization of data
% create a for loop to apply the normalization
formula to each of the input and output data. For
example;

for i = 1 : length(Input_Data)
Normalised_Input_Data = normalization formula;
Normalised_Input_Data = normalization formula;
end

X = Normalised_Input_Data;
Y = Normalised_Input_Data;

%% 80:20 training and testing
x_train = 80% of X;
y_train = 80% of Y;
x_test = remaining 20% of X;
y_test = remaining 20% of Y;

%% Model Development
rng default;
Mdl = fitrsvm/fitcsvm(x_train, y_train,
'KernelFunction', 'function name', 'KernelScale',
'positive scalar', 'Standardize', 'logical',...Name,
value);
```

```
%% Testing the developed Model

y_predicted_test = predict(Mdl, x_test);
error_test = (y_test - y_predicted_test);
mean_absolute_error = mae(error_test);
root_mean_square_error = rms(error_test);
r_value = corrcoef(y_test,y_predicted_test);
```

```
%% De-normalising and testing the Model
```

% create a for loop to apply the de-normalization
formula to each of the predicted output data and
then find out the statistical parameters again like
RMSE, MAE, r.

```
%% Plotting the scattered and regression line

plot(x);
hold on;
plot(y);
```

```
%% save all the results

save filename.mat;
```

```
%% export the results to files

xlswrite('filename');
```

- **Artificial Neural Network (ANN)**

```
  %% clear workspace and command window

  clear;
  clc;
  close all;
```

```
  %% prepare dataset

  Input_Data = xlsread(filename);
  Output_Data = xlsread(filename);
```

```
  %% Normalization of data
```

  % create a for loop to apply the normalization
  formula to each of the input and output data. For
  example;

```
  for i = 1 : length(Input Data)
  Normalised_Input_Data = normalization formula;
```

```
Normalised_Input_Data = normalization formula;
end

X = Normalised_Input_Data;
Y = Normalised_Input_Data;
```

80:20 training and testing

```
x_train = 80% of X;
y_train = 80% of Y;
x_test = remaining 20% of X;
y_test = remaining 20% of Y;
```

Defining Model parameters

```
Training_function = function_name;
Hidden_Layer_Size = Integer;
```

% Try different number of hidden layer size and record the results. Finally select the one for which optimum value is achived.

Model Development

```
rng default;
net = fitnet(Hidden_Layer_Size,Training_Function);
[net,tr] = train(net,x_train',y_train');
```

Testing the developed Model

```
y_predicted_test = net (x_test');
error_test = (y_test - y_predicted_test);
mean_absolute_error = mae(error_test);
root_mean_square_error = rms(error_test);
r_value = corrcoef(y_test,y_predicted_test);
```

De-normalising and testing the Model

% create a for loop to apply the de-normalization formula to each of the predicted output data and then find out the statistical parameters again like RMSE, MAE, r.

Plotting the scattered and regression line

```
plot(x);
hold on;
plot(y);
```

```
%% save all the results
save filename.mat;
%% export the results to files
xlswrite('filename');
```

- **Adaptive Neuro-Fuzzy Inference System (ANFIS)**

```
%% clear workspace and command window
clear;
clc;
close all;

%% prepare dataset
Input_Data = xlsread(filename);
Output_Data = xlsread(filename);

%% Normalization of data
% create a for loop to apply the normalization
formula to each of the input and output data. For
example;

for i = 1 : length(Input_Data)
Normalised_Input_Data = normalization formula;
Normalised_Input_Data = normalization formula;
end

X = Normalised_Input_Data;
Y = Normalised_Input_Data;

%% 80:20 training and testing
x_train = 80% of X;
y_train = 80% of Y;
x_test = remaining 20% of X;
y_test = remaining 20% of Y;

%% Selection of FIS Generation Method
Option{1}='Grid Partitioning (genfis1)';
Option{2}='Subtractive Clustering (genfis2)';
Option{3}='FCM (genfis3)';

%% Setting the Parameters of FIS Generation Methods
For Option{1}: {'Number of MFs', 'Input MF Type',
'Output MF Type:'};
For Option{2}: {'Influence Radius:'};
```

For Option{3}:{'Number of Clusters:', 'Partition Matrix Exponent:','Maximum Number of Iterations:', 'Minimum Improvement:'};

%% Model Development

```
rng default;
fis=genfis([x_train, y_train],...Name, Value);
fis=anfis([x_train y_train], fis, Train_Options,
Optimization_Method);
```

%% Testing the developed Model

```
y_predicted_test = evalfis (x_test,fis);
error_test = (y_test - y_predicted_test);
mean_absolute_error = mae(error_test);
root_mean_square_error = rms(error_test);
r_value = corrcoef(y_test,y_predicted_test);
```

%% De-normalising and testing the Model

% create a for loop to apply the de-normalization formula to each of the predicted output data and then find out the statistical parameters again like RMSE, MAE, r.

%% Plotting the scattered and regression line

```
plot(x);
hold on;
plot(y);
```

%% save all the results

```
save filename.mat;
```

%% export the results to files

```
xlswrite('filename');
```

Index